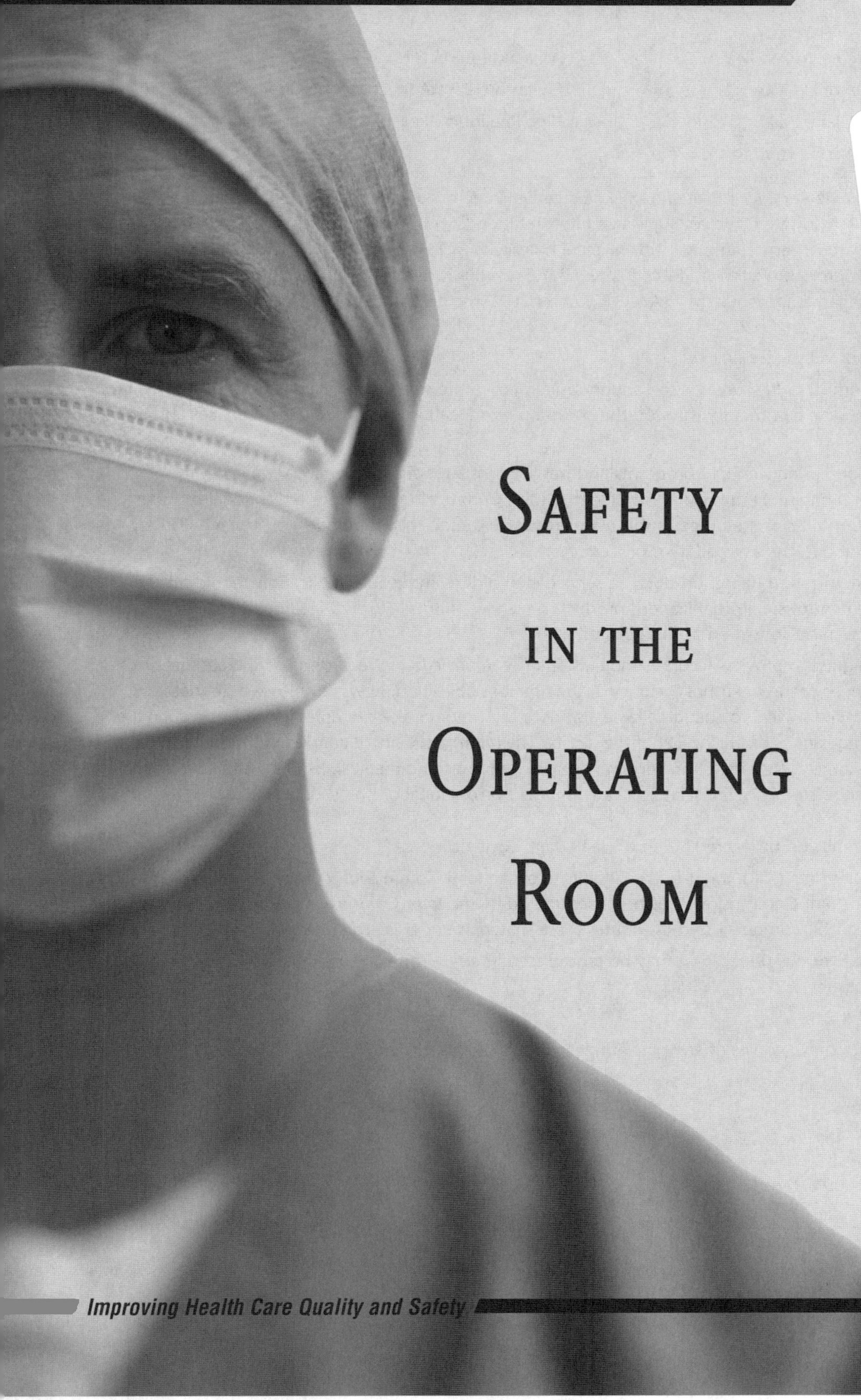

Joint Commission
RESOURCES

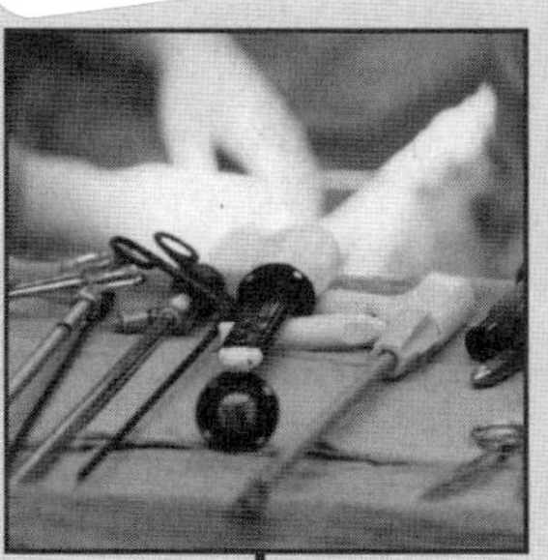

Safety in the Operating Room

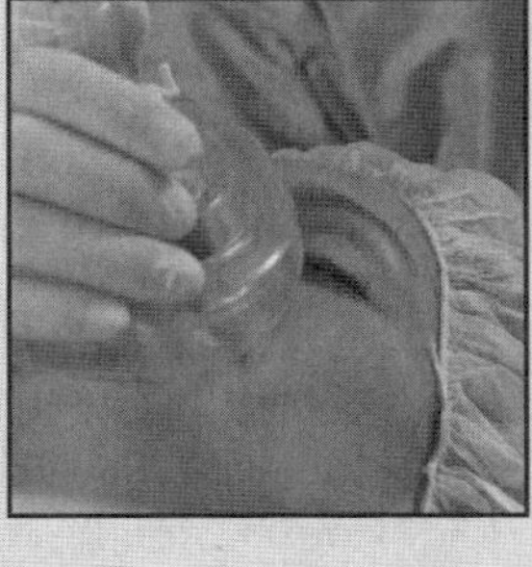

Improving Health Care Quality and Safety

Executive Editor: Kristine M. Miller, M.F.A.
Project Manager: Jan Kendrick, M.A.
Manager, Publications: Diane Bell
Production Manager: Johanna Harris
Associate Director: Cecily Pew
Executive Director: Catherine Chopp Hinckley, Ph.D.
Vice President, Learning: Charles Macfarlane, F.A.C.H.E.

Joint Commission/JCR Reviewers: Rick Croteau, M.D., Executive Director for Patient Safety Initiatives; Jerry Gervais, C.H.F.M, C.H.S.P, Associate Director, Standards Interpretation Group; Louise Kuhny, R.N., M.P.H., C.I.C., Associate Director, Standards Interpretation Group; Michael Kulczycki, M.B.A., Executive Director, Ambulatory Care Accreditation Program; Laurel Ann Peterson, R.N., B.S.N., Standards Interpretation Specialist, Standards Interpretation Group; Megan E. Sawchuk, M.T.(A.S.C.P.), Associate Director, Standards Interpretation Group

Joint Commission Resources Mission

The mission of Joint Commission Resources is to continuously improve the safety and quality of care in the United States and in the international community through the provision of education and consultation services and international accreditation.

Joint Commission Resources educational programs and publications support, but are separate from, the accreditation activities of the Joint Commission. Attendees at Joint Commission Resources educational programs and purchasers of Joint Commission Resources publications receive no special consideration or treatment in, or confidential information about, the accreditation process.

The inclusion of an organization name, product, or service in a Joint Commission publication should not be construed as an endorsement of such organization, product, or service, nor is failure to include an organization name, product, or service to be construed as disapproval.

This publication is designed to provide accurate and authoritative information in regard to the subject matter covered. Every attempt has been made to ensure accuracy at the time of publication; however, please note that laws, regulations, and standards are subject to change. Please also note that some of the examples in this publication are specific to the laws and regulations of the locality of the facility. The information and examples in this publication are provided with the understanding that the publisher is not engaged in providing medical, legal, or other professional advice. If any such assistance is desired, the services of a competent professional person should be sought.

Joint Commission Resources, Inc. (JCR), a not-for-profit affiliate of the Joint Commission on Accreditation of Healthcare Organizations (Joint Commission), has been designated by the Joint Commission to publish publications and multimedia products. JCR reproduces and distributes these materials under license from the Joint Commission.

Printed in the U.S.A. 5 4 3 2 1

Permissions Editor
Department of Publications
Joint Commission Resources
One Renaissance Boulevard
Oakbrook Terrace, Illinois 60181
permissions@jcaho.org

ISBN: 0-86688-998-1

Library of Congress Control Number: 2006931085

For more information about Joint Commission Resources, please visit http://www.jcrinc.com.

TABLE OF CONTENTS

Introduction

A man has surgery on the wrong arm. A child receives the wrong medication in the operating room (OR). A woman is burned when her surgical drape catches fire during a procedure.

All these events can and do happen in an OR, and most of them have devastating consequences. Patients are harmed, health care organizations are sued, staff members are psychologically scarred, and the media reports on it all with fervent attention to detail.

Nowhere, perhaps, is it more important to preserve the safety of patients than in the OR. Lives often depend on it. While the consequences of errors in the OR may be obvious, the causes of these errors may be more subtle and include distractions in the OR environment, stress of the perioperative staff, failures in communication, and lack of training and education.

Safety in the Operating Room discusses a variety of safety concerns faced in the perioperative environment. It offers suggestions on how organization leadership, perioperative staff, and patients can work together to address these concerns. In this book, each chapter starts with a hypothetical story about an issue involving the chapter topic. This hypothetical story is meant to provide context for the topic and illustrate its human aspect. Within the chapters, relevant Joint Commission standards and National Patient Safety Goals are offered for reference. (For a complete list of Joint Commission standards and National Patient Safety Goals, organizations should consult the accreditation manuals for their programs.)

Content of the Book

Chapter 1 of the book discusses the importance of effective communication. Nearly all the safety risks covered in this book can be reduced through effective communication. It is the underpinning of a safe perioperative environment and a first step in improving safety in the OR.

The rest of the book is divided into chapters discussing patient-associated, OR–associated, and surgery-associated risks. Chapters 2 through 10 address the following possible safety concerns:

- Wrong-site, wrong-procedure, wrong-patient surgery
- Medication errors
- Surgical fires
- Infection control issues
- Hand-hygiene issues
- Sharps injuries
- Overbooking of ORs
- Anesthesia awareness
- Retention of foreign bodies

Two appendixes deal with environment of care and impaired provider concerns in the OR.

While the focus of this book is on the OR, the entire perioperative environment is considered. This includes the preoperative area, the OR itself, and the postoperative care area. Discussions are relevant to traditional ORs, such as those found in hospitals, as well as ambulatory surgery units, cardiac catheterization suites, endoscopy suites, and other areas where surgical and other invasive procedures may be performed.

Using a Proactive Approach to Performance Improvement

Most of the errors discussed in this book can be prevented. Organizations should consider taking a proactive approach to addressing safety issues by examining possible opportunities for error and designing and implementing risk-reduction strategies. There are several methods organizations can use to proactively address safety risks. Failure mode and effects analysis (FMEA) and root cause analysis (RCA) are just two of the tools that can be used for this purpose. Organizations wishing to learn more about proactive risk assessment should go to the Joint Commission's Web site, at http://www.jointcommission.org.

Not all suggestions offered in this book will be relevant or appropriate for every organization. For example, suggestions that are appropriate for a hospital setting may not be effective in an ambulatory surgery center. Organizations should use this book to consider the important safety problems and solutions and relate them to their particular settings.

Audience for the Book

Surgical teams, including nurses, surgical technologists, surgeons, and anesthesiologists, can benefit from reading this book. However, the audience is not limited to perioperative staff and also includes clinical leaders, medical leaders, patient safety officers, risk managers, staff educators, pharmacists, facilities managers, and others responsible for patient and environmental safety in the OR.

Safety in the OR requires multidisciplinary involvement. Nurses, surgical technologists, surgeons, anesthesiologists, facility managers, engineers, organization leaders, and patients must all work together to provide the safest possible care. The risks discussed in this book are manageable. The OR can be a safe place. Organizations that empower staff members to work together to implement and consistently follow processes, protocols, and policies to address these risks can achieve a safe and quality OR environment, where the needs of patients and their families are met and their expectations are exceeded.

Acknowledgements

A work of this scope requires a concerted team effort. This volume, which we hope you find to be as useful as it is practical, could not have been created without such a team. We are deeply indebted to our writer, Kathleen B. Vega, for her hours of researching, interviewing, and writing; her careful attention to detail; and her consummate professionalism.

We wish to acknowledge the outstanding contribution to the completion of this book by our Joint Commission reviewers (see the masthead on page ii) and by the following outside expert reviewers:

- Pauline Robitaille, R.N., M.S.N., C.N.O.R., Former Chief Nursing Officer/Vice President, Professional Services, Association of periOperative Registered Nurses, Denver
- F. Dean Griffen, M.D., Chairman of the Patient Safety and Professional Liability Committee, American College of Surgeons; Clinical Professor of Surgery at Louisiana State University Health Sciences Center, Shreveport, LA
- John A. Schaefer, M.F.S., C.I.H., H.E.M., C.P.E.A., Associate Director, Health, Safety and Environment; Assistant Professor of Medicine/Environmental Health Sciences, the Johns Hopkins University, Baltimore

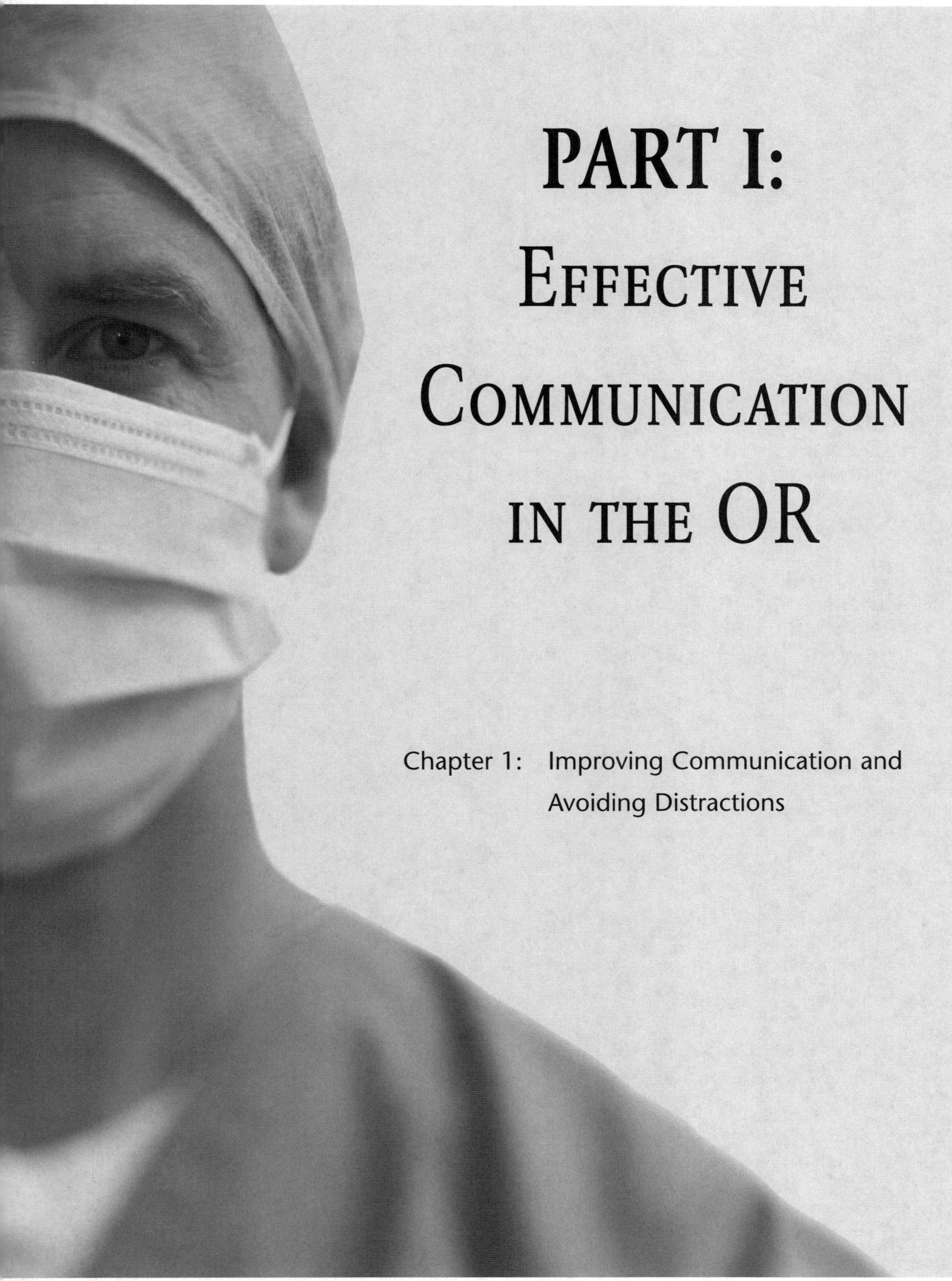

PART I: Effective Communication in the OR

Chapter 1: Improving Communication and Avoiding Distractions

It Could Have Happened This Way

Carrie Fitzgerald was five years old and needed a complex and somewhat rare procedure to save her life. A surgeon who was experienced but new to the hospital was scheduled to lead the surgery. To be successful, all members of the surgical team, including the surgeon, needed to work together and coordinate the surgery.*

On the day of the surgery, Dr. Gregory Brown walked into the operating room (OR) and saw the anesthesiologist, circulating nurse, scrub nurse, and orderly all working diligently to prepare for surgery. As Dr. Brown entered, everyone turned to look at him. "Good morning," he said. "My name is Greg Brown. I hope we can work together and help this patient. I have no pride invested in this procedure; I just want us to get it right. If you see anything that is helpful or see me starting to make a mistake, please speak up." Dr. Brown then went around the room and introduced himself to everyone individually, asking them to call him Greg. The team held a briefing before the start of the procedure; conducted a time-out to confirm the correct site, patient, and procedure; and identified some possible risk areas and brainstormed solutions. During critical components of the procedure, Dr. Brown asked for and received silence. At one point during the procedure, the circulating nurse determined that there was a count discrepancy. She alerted Dr. Brown, who halted the procedure while everyone looked for the missing sponge. Once they found it, the procedure resumed. "Glad you caught that," said Dr. Brown. "Nicely done."

The perioperative team operated on Carrie for five hours and successfully completed the procedure. At the end, Dr. Brown thanked everyone by name for his or her work and communicated a hope that they would all work together again soon.

Carrie recovered normally and showed no symptoms of an infection or adverse effects. Dr. Brown continued to practice at the hospital, and the other members of the surgical team always looked forward to working with him. They all knew that he was a competent professional who cared about his patients, respected all members of the surgical team, and valued everyone's input. In addition, the hospital rarely had to spend unanticipated time and resources treating his patients.

This example exhibits a standard of what communication in the OR should be: all team members working together, following protocols, respecting each other, and valuing each other's opinions. The results of such communication include improved patient outcomes; enhanced staff satisfaction; and improved financial outcomes, as organizations do not need to spend resources–such as time, money, supplies, and staff–on addressing problems after they have occurred.

Why Effective Communication Is Important in the OR

Perioperative teams regularly deal with the challenges of sophisticated instruments, the rapid transfer of critical information, the changing nature of a patient's condition, and the uncertainty inherent in surgery.[1] At the same time, perioperative team

* All patient and care-provider names in this book are fictitious.

members may know little about each other or lack a common understanding of the needs of a particular patient or procedure. Consequently, when the perioperative staff does not communicate effectively, critical information can be lost, misunderstood, or forgotten, thus leading to lapses in patient safety. Communication failure is the root cause of nearly 70% of sentinel events* reported to the Joint Commission, making it the most frequently cited root cause. A significant portion of these sentinel events have resulted in patient death.[2]

Within most operations, a variety of protocols, processes, and procedures must be carried out effectively and efficiently. This book discusses many such examples, including conducting a time-out, counting supplies, confirming medications, and consistently monitoring anesthesia delivery. Effective communication is the glue that holds all these protocols, procedures, and processes together to ensure a comprehensive and safe process.

Creating a Culture of Communication

Effective communication is complete, accurate, timely, unambiguous, and clearly understood. Joint Commission standards require organization leadership to set the tone for effective communication throughout the organization. Effective communication cannot take place in the OR if an organization's culture does not value, support, and require open and honest communication. Some of the ways to establish and support a culture of communication include providing the staff access to leadership, eliminating hierarchies between staff members, and encouraging a team approach to care.

Standard LD.3.60. Communication is effective throughout the organization.

Providing the Staff Access to Leadership

Perioperative staff should have access to organization leadership to provide general feedback, express concerns and frustrations, and celebrate successes. Not only will staff members feel their opinions are valued, but leadership can learn from this information and determine areas for improvement. Staff surveys, focus groups, suggestion boxes, and issue hotlines are all effective ways of listening to staff feedback, to which leaders should respond in a timely manner. If a staff member expresses an opinion or a concern and receives no response from leadership, the staff member may feel ignored and undervalued and may be less likely to provide feedback in the future. One way to solicit staff feedback is through a reporting system. This is a communication venue in which leadership can learn about medical errors, near misses, and causes of concern. More information on reporting systems can be found in Chapter 3.

Eliminating Hierarchy

A culture of communication emphasizes a team approach to care in which individuals work together, share knowledge, and coordinate efforts to provide the optimum level of care. Effective teams do not allow hierarchy to be present. When hierarchy differences exist, people on the lower end of the hierarchy tend to be uncomfortable, if not downright afraid, to speak up about problems or concerns. Gruff, aggressive, or intimidating behavior by individuals at the top of a hierarchy can stifle communication and give the impression that such individuals are unapproachable. Conversely, individuals on the higher end of the hierarchy are

* The Joint Commission defines *sentinel event* as an unexpected occurrence involving death or serious physical or psychological injury, or the risk thereof. Serious injury specifically includes loss of limb or limb function. The phrase *or the risk thereof* includes any process variation for which a recurrence would carry a significant chance of a serious adverse outcome. Further descriptions of sentinel events can be found on the Joint Commission Web site, http://www.jointcommission.org.

more likely to think that concerns brought up by others are unwarranted or challenging to their authority. Consider this hypothetical example:

A scrub nurse at a large community hospital watches an anesthesiologist numb the wrong shoulder of a patient who is having shoulder surgery. The nurse desperately wants to tell the physician he is making an error, but this physician has a history of reacting poorly to correction and retaliating against those who question him. The nurse keeps her mouth shut, and the patient is prepped for surgery on the wrong arm.

Organizations that want to eliminate hierarchy in the OR must establish such a climate across the organization. Organization leadership must encourage collaboration, respect the input of all staff, and accept nothing less than team performance. Organizations should consider using structured communication techniques that eliminate hierarchy, such as briefings and assertive conversation checklists (see pages 6–9).

Physicians, in particular, should be encouraged to eliminate hierarchies. Often, physicians are not aware of hierarchies that exist between them and other staff members because they are usually at the top of any hierarchy. Thus, physicians may perceive the health care environment as collaborative when it really is not. Providing feedback to physicians on their interaction with other staff members may be helpful in alerting them to hierarchy issues and empowering them to change (see page 10).

☞ **Tip:** In addition to eliminating hierarchies on an organizational level, perioperative team leaders must also work to eliminate hierarchies within the perioperative team. While effective teams do not have a hierarchy, they do have a leader. This may seem like a subtle difference, but it is a very distinct one. A team leader sets the tone for perioperative team interactions. He or she works to flatten hierarchies by using first names, engaging team members in the care process, and incorporating the suggestions and expertise of individual team members into the clinical plan.[3] To eliminate hierarchy, the team leader must ensure that all participants are treated respectfully, that everyone's input is valued, and that the team remains focused on the task at hand. In many cases, surgeons are the perioperative team leaders; however, this is not always the case. An anesthesiologist, a nurse, a respiratory therapist, or another clinician may take a leadership role if the action the team is performing is in his or her area of expertise. Team leaders should receive special training on how to lead an effective team in a manner that eliminates hierarchy and encourages collaboration. ☜

To eliminate hierarchies between staff members, an organization may need to overcome the educational experiences of its staff. Traditionally, health care providers, particularly physicians, are trained to think as individuals rather than as team members. Little emphasis has been placed on how to talk with different disciplines about patient care or respond when orders or decisions are questioned. Communication skills such as collaboration, debriefing, and conflict resolution are not taught in many medical and nursing schools around the country. To overcome this knowledge gap, organizations should consider teaching communication skills to all health care providers and providing team training to perioperative staff.

Encouraging a Team Approach to Care

Effective teams are characterized by trust, respect, and collaboration. Team members value familiarity over formality and watch out for each other to avoid mistakes. To encourage a team approach to care, organizations may take a variety of actions:

- Using structured communication tools that eliminate hierarchies
- Encouraging feedback
- Providing team training
- Addressing disruptive behavior

> **Sidebar 1-1. Looking to Aviation for Answers**
>
> There are a variety of ways organizations can encourage a team approach to care. As no organization has time to reinvent the wheel, it may be helpful to look toward other industries, such as aviation, for examples of how to encourage teamwork and communication.
>
> In the late 1970s, the aviation industry began addressing the issue of communication among its flight crews. The industry discovered that within these crews, hierarchies existed, individuals felt uncomfortable challenging pilots, and interactive communication was almost nonexistent. In many ways, the crew of a large aircraft was similar to a surgical team in an OR. Historically, the surgeon, like the pilot, assumed absolute control and responded poorly to questions about his or her decisions.
>
> To overcome communication problems, the aviation industry developed a training program that added teamwork skills to the traditionally taught technical skills. Crew resource management (CRM) is now widely used by airlines, and it helps teach the flight crew how to draw upon the different capabilities and knowledge of a given crew. Assertiveness training, situation awareness, adaptability, leadership, and decision making are all taught as a part of the program. Crew members engage in simulation exercises, role-playing exercises, and self-critiques to practice effective communication techniques. Interactive group debriefings are also used to help improve communication.[4]

The following sections discuss these actions and provide suggestions on how organizations can effectively achieve a team approach to care.

Using Structured Communication Techniques

When a team must communicate complex information in a short period of time and miscommunication or lack of communication can have serious consequences, it can be helpful to use structured communication techniques. These prescriptive steps ensure that the proper information is conveyed at the correct time to the correct people. Just like clinical practice guidelines, structured communication techniques may assist practitioners in making decisions and taking actions for specific clinical circumstances.[5] Following are some of the most common structured communication techniques.

OR Briefings. Briefings are short discussions between team members that aim to position all clinicians at the same starting point. Briefings can help promote a sense of collaboration between team members, set the tone for open communication, and bring everyone onto the same page to avoid surprises. During briefings, perioperative team members can discuss the operative plan and what they want to accomplish during surgery. They can identify patient risks, potential hazards, and safety concerns. They can compare notes, share knowledge, and identify any needed resources.

Briefings can be used at any point during surgery. For example, briefings can set the stage for the process or can help a team refocus during times of change, such as a sharp decline in patient status. Successful briefings are structured and quick. They involve all team members. Individuals conducting a briefing should use first names and make eye contact with other team members. These strategies establish a tone of collaboration and encourage individuals to speak up.[3]

To help navigate the briefing process, a team of quality and safety researchers at the Johns Hopkins Medical Institutions created an OR briefing tool that includes instructions and a checklist. (See Figures 1-1 and 1-2.) The purpose of the tool is to provide a framework for communicating information that is necessary for all team members to participate effectively. In addition, it serves as a memory and efficiency tool to help avoid communication failures. The tool allows the staff to express concerns, identify hazards, and plan ways to address those hazards.[6]

Debriefings. While a briefing can be helpful before or during a procedure or shift change, a *debriefing* can be helpful after a process is complete. A debriefing allows team members to discuss how things went, the outcomes, the opportunities for improvement, and the elements of success. Complications, surprises, and unanticipated events–such as equipment failures–can be identified and addressed during this time.[7] The debriefing process allows a team to better prepare for the next encounter and avoid repeating the same mistakes.

Situational Awareness. Situational awareness occurs when all team members agree about a procedure or process under way, recognize potential problem areas, and know the plans to address these areas. An ongoing dialogue between team members is required to maintain situational awareness. If situational awareness is lost, communication can break down, and mistakes can occur. The following are some red flags that signal a loss of situational awareness:

- Ambiguity
- Reduced communication
- Confusion
- Trying something new under pressure
- Deviating from established norms
- Verbal violence
- Things not "feeling right"
- Fixation
- Boredom
- Task saturation
- Being rushed or behind schedule[3]

To help maintain situational awareness in the OR, surgical teams can use a variety of communication strategies, including the following:

- *Callout.* This communication technique requires team members to call out each phase of a process as they are entering it so that all team members know it is safe to proceed. In the OR, callouts are used during at least two points: the start and the

OR Briefing Tool–User Instructions

Problem Statement

Communication failures have been the root causes in OR sentinel events, wrong-site surgery, and other medical errors in the OR. Effective communication must be a primary component of any team-based activity, particularly surgical procedures.

OR Teams

OR teams are composed of a group of multidisciplinary providers who may or may not be familiar with one another. In addition, each provider has discipline-specific priorities and constraints, such that any single team member is rarely aware or knowledgeable of important details involved in the case before arriving in the OR.

Purpose of Tool

The purpose of this tool is to provide a framework for communicating the necessary information that all OR team members need to participate as effective members of the team and ensure a safe, successful procedure.

Who Should Use This Tool

Any member of the surgical team, often the surgeon, can initiate the briefing; however, representatives from surgery, anesthesiology, and nursing each have relevant information to share during this discussion.

When to Use This Tool

This tool should be employed when all members of the team are in the OR and after the patient is anesthetized.

Figure 1-1. *The Johns Hopkins Medical Institutions created this tool to help navigate the briefing process. The checklist portion of the tool is in Figure 1-2.*

Source: Joint Commission Resources, *Joint Commission Journal on Quality and Patient Safety,* June 2006, p. 352, with permission of the Johns Hopkins Medical Institutions.

OR Briefing Checklist

This checklist helps ensure consistent and complete communication in the OR. The surgeon leads Part I, while all team members identify relevant information in Sections II and III.

Implementation Instructions. Make sure all team members are in the OR suite and the patient is anesthetized. Just prior to starting the procedure, implement the checklist, beginning with the introduction of names, roles, and work, in descending fashion, through the list.

I. Introduction of first names and roles, which are written on the whiteboard

II. Review critical information

- ❑ Do we have the correct patient?
- ❑ Is the correct side or site marked?
- ❑ Has the procedure been determined?
- ❑ Have antibiotics been given?

III. Identify and mitigate hazards

SURGERY: Discuss plans for the surgical procedure:

- ❑ Describe critical steps
- ❑ Provide team with pertinent information, including problems that may be encountered
- ❑ Ask team: If something were to go wrong with this procedure, what would it be, and how could we prevent the problem?
    - ❍ Risks during procedure, such as bleeding, fluid loss
    - ❍ Surgeon suggests, "If anyone has a concern during the case, please let me know."

ANESTHESIOLOGY: Discuss all relevant issues:

- ❑ Patient comorbid disease that will increase risk
- ❑ Aspects of surgery that increase risk, such as need for IV access
- ❑ Availability of blood products
- ❑ Interventions to prevent complication, such as myocardial infarction, surgical site infection

NURSING: Discuss all relevant issues:

- ❑ Are all necessary instruments available?
- ❑ Will any special equipment be considered?
- ❑ Plan for breaks (relieving nurse to introduce himself or herself when switching)

Figure 1-2. *The Johns Hopkins Medical Institutions created this tool to help navigate the briefing process. The instruictions for using the tool are in Figure 1-1.*

Source: Joint Commission Resources, *Joint Commission Journal on Quality and Safety,* June 2006, p. 353, with permission of the Johns Hopkins Medical Institutions.

close of a procedure. Surgical teams may also use the callout technique for other reasons, such as to indicate that the sponge count is correct or that the patient is coming off bypass. When using the callout technique, participants should speak clearly and loudly so all team members can hear.[7]

■ ***Readback/hearback.*** This is a closed-loop communication tool that helps improve the reliability of an information exchange between two or more people.[8] The tool involves four distinct actions:

1. The "sender" concisely states information to the "receiver."

2. The receiver repeats back what he or she heard.

3. The sender then acknowledges that the repeatback was correct or makes a correction.

4. The process continues until a shared understanding is verified.

Within this model, responding to a message with "okay" or "uh-huh" is not sufficient to close the communication loop. The message must be explicitly restated and acknowledged. Organizations that require this type of closed-loop communication during surgery help smooth the communication process and ensure that no critical information is lost. If perioperative team members practice this type of communication during routine surgical circumstances, then they will be well prepared to use it during acute instances when time pressure and urgency play a role.[8] The readback/hearback approach can be used successfully in communicating verbal medication orders. More information on this can be found in Chapter 3.

Assertive Conversation Checklist. Although the presence of hierarchies can intimidate individuals and prevent them from vocalizing concerns or problems, hierarchy is not the only reason people do not speak up. Sometimes individuals simply do not feel comfortable sharing bad news or asserting themselves. When people assert themselves, they should be pleasant yet persistent. Being assertive does not mean an individual should be aggressive, hostile, or confrontational. Following are five elements to making an assertive statement:

1. Secure the individual's attention
2. State the concern
3. State the problem
4. Offer a solution
5. Obtain agreement[7]

In some cases, providing individuals with an assertion checklist or model can be helpful. This list indicates steps individuals can take to effectively and positively assert themselves so that their opinion is heard. The following is an example of such a checklist:

1. Secure the person's attention. Using his or her name can help gain notice.

2. Make eye contact and face the person.

3. Express your concern.

4. State the problem clearly and concisely. Rambling or speaking softly can indicate that you are unsure of your point, and thus your point may be disregarded.

5. Propose an action.

6. Make sure that the team understands the problem and proposed action.

7. Reach a decision.

8. Affirm that all team members understand the decision.

9. Reassert, if necessary.[3]

The Sterile Team Environment. At certain points during surgery, teams may wish to cease all discussion and outside interruptions. By asking for a "sterile team environment," the team leader indicates that the procedure is entering a critical phase

Sidebar 1-2. A Chain-of-Command Policy

It is often difficult for a health care worker to challenge an order given by a physician. There are times, however, when challenging physician orders is appropriate, and even necessary, to prevent an error. Organization leaders might want to establish a chain-of-command policy to help the staff feel more comfortable questioning orders. A chain-of-command policy is a formal administrative tool that provides employees with a process to resolve administrative, clinical, or other patient safety or service issues. Employees can use a chain-of-command policy and procedures to present an issue of concern and pass it up the lines of authority until it is resolved.[9] All staff members should be trained on the policy and assured that evoking it will not be held against them.

and that the team should talk only about the task at hand. Flight crews have used this technique during takeoffs and landings. The purpose is to avoid distractions that may lead to errors.[7] Evoking a sterile team environment may be appropriate during critical procedures; during sponge, sharp, or instrument counts; when verifying the surgical site, procedure, and patient; and any other time the full attention of the surgical team is needed.

PROVIDING FEEDBACK

If a physician in most health care organizations is asked if he or she communicates well, the answer will probably be "of course." However, if a nurse is asked whether that physician communicates well, he or she may have a different answer. Many times physicians and other organization staff are not aware of any shortfalls in their communication skills. That is why it is important to provide them with regular and constructive feedback.

Organizations can gather information for feedback from a multitude of sources, including patient and peer surveys, focus groups, direct observation, and patient and staff complaints. A recent study by Violato and his colleagues described a program that assessed surgeon performance through multisource feedback from questionnaire data. The process assessed a variety of performance areas, including communication skills, interpersonal skills, diagnostic and treatment skills, collaboration skills, professionalism, support for colleagues, accessibility, stress-management skills, and a demonstrated ability to continuously improve. The data came from patients; medical colleagues; and coworkers, including nurses, pharmacists, and psychologists; and the surgeons themselves. To collect data, a committee of surgeons from major surgical disciplines developed questionnaires for physician colleagues, coworkers, and patients. Each group had its own survey, with comparable data points. Participants in the survey were asked to rate a surgeon on a five-point scale. The instrument summarized the data by item and category and compared it to the physician's specialty group.[10]

Regardless of how an organization gathers feedback on a physician's ability to communicate, it is important to share that information. This can help encourage physicians to participate in communication initiatives, improve their practice, and incorporate principles of good communication into their daily work habits.

PROVIDING TEAM TRAINING

Merely telling OR professionals to "act as a team" will achieve few, if any, results. An important step in creating effective teams is to provide team training to all staff members that teaches them how to be a part of an effective team. Without training, teams are just highly diverse groups of people that work side by side. Training can mold a group into a unified entity that can face problems, identify issues, and work together to care for a patient.

Perioperative team training should address the value of a team approach to care and also teach staff members how to function as a team. Education efforts should include the importance of body language, how to resolve conflicts, and how to communicate effectively yet respectfully. Training may involve bringing groups of perioperative staff members together to get to know each other, engage in team-building exercises, talk frankly about current communication issues, and work together to solve communication problems. Team training can also involve role-playing exercises, which can build awareness of one's own communication patterns and provide structured opportunities for the practice and critique of new behaviors.[11]

☞ **TIP:** Some organizations choose to conduct perioperative team training exercises off site during staff retreats. This strategy places special emphasis on the training and communicates to the staff that teams are an important part of the organization's culture. An organization might

Sidebar 1-3. Ten Ways to Communicate Effectively

The following list offers some ways team members can communicate effectively. Organizations may consider incorporating these suggestions into team training content.

1. Listen without judgment. The key to good communication is listening well. Individuals should save their judgments for after they have heard and understood what was said.
2. Listen with the willingness to be swayed by the other person's opinions. Individuals are under no obligation to be swayed, but they should stay open to all options.
3. Listen without thinking about what to say next. Individuals should take time before they respond.
4. Do not be invested in being right. Being right is not the point. If a person must be right, then he or she is not able to listen or communicate because a barrier already exists. If a person is always right, then other people are always wrong, which simply cannot be true.
5. Ask for repetition if one's mind wanders. It is human nature to be distracted.
6. Repeat back what was heard and ask if it is correct.
7. Pay attention to body language.
8. Be honest but consider the listener's feelings. People can be truthful yet be polite, respectful, and sincere.
9. Get comfortable with gray. Most things are not black and white but somewhere in the middle.
10. Have integrity and build trust. Individuals should not say what they do not mean. They should not make promises they will not or cannot keep. They should follow through with any commitments they make.

Source: Adapted from Lehavi D.: *Top Ten Ways to Improve Your Communication Skills.* SolveYourProblem.com, Jul. 19, 2004. http://www.solveyourproblem.com/artman/publish/printer_387.shtml (accessed Jun. 21, 2006).

want to bring in an outside training company to work with staff on team training, depending on the type and size of the organization. ☜

Despite the importance of team training, many organizations do not engage in this activity. As one expert put it, "It is ironic indeed to realize that a football team spends 40 hours a week practicing teamwork for the 2 hours on Sunday afternoon when teamwork really counts. Teams in organizations seldom spend 2 hours per year practicing, when their ability to function as a team counts 40 hours per week."[12]

☞ **Tip:** In most organizations, physicians are at the top of any perceived hierarchy. Because of the high-profile nature of their positions, physicians must support, endorse, and actively participate in perioperative team training and team building in order for successful team-based care to take place. Physicians with surgical and anesthesia privileges must be involved in any training exercises. Organizations may want to identify some champions within the physician community who can actively promote, support, and participate in the process of team training. Peer pressure can be a very good motivator in encouraging other physicians to be part of a team. ☜

Discussing the Importance of Nonverbal Communication. While the spoken word is important, it is not the only way in which messages are conveyed.[14] Tone, eye gaze, stance, facial expression, and hand gestures can communicate their own messages, which may or may not be synchronized with the words that are being said. Nonverbal cues can be very powerful. For example, if a nurse alerts a surgeon to a discrepancy in the sponge count, and the surgeon reacts by sighing and rolling his eyes, he is communicating annoyance and irritation, even if his words do not.

It is estimated that as much as 93% of communication includes nonverbal behaviors such as tone of voice, mannerisms, and body language.[14] When providing team training, special attention should

Sidebar 1-4. Sharing Stories

Everybody has a story of a procedure that went particularly well or particularly poorly. In most cases, individuals have learned from those experiences. How often has someone said, "I learned my lesson. I'm never doing that again." Team training for perioperative staff can be enhanced by storytelling because it teaches others about mistakes and successes. Organizations should encourage staff members to share their stories so that everyone can learn from them. Consider the following example:

> A nurse goes into the preoperative holding area to retrieve her patient, and the patient is not there. Panicked, the nurse looks in each operating suite. She finds her patient in the wrong room, being prepped for the wrong surgery. She stops the prep process and gets her patient to the right room. The nurse shares this story with other staff members, who are relieved that the error was prevented. The story reminds everyone who hears it about the importance of checking the patient's wristband before surgery.[13]

To help collect staff stories, the Association of periOperative Registered Nurses (AORN) has developed SafetyNet, a near-miss reporting system. This system collects stories for perioperative nurses and clinicians about occurrences that almost happened. AORN analyzes these stories, learns about common themes, and offers practice guidance to prevent similar events from occurring. To contribute a story to SafetyNet, organizations can go to http://patientsafetyfirst.org/safetynet.html.

be paid to nonverbal communication. In other words, teams must learn that how they communicate is just as important as what they say.

Offering Cultural Training. In most organizations, the staff comes from a variety of cultural backgrounds. The United States is one of the most racially, ethnically, and culturally diverse countries in the world. In all interactions, cultural differences can exacerbate communication problems. For example, the interaction of male Ethiopian surgeons with female nurses may be very challenging. Because of their national enculturation, these physicians may hesitate to grant expertise or authority to female nurses. Within the Asian culture, individuals may be reluctant to be assertive and challenge opinions openly because loss of face is a major issue. As a result, it may be very difficult for Asian nurses to speak up if they see something worrisome. They often communicate their concern in very indirect ways.[3]

Making sure that communication efforts respect individuals' cultural differences and overcome any barriers that may be caused by these differences is paramount to achieving effective communication. Cultural training for perioperative staff is critical to this effort. Staff members should be aware of the cultural makeup of the organization and understand the attitudes and beliefs of the people with whom they work. Leaders should make sure that any procedures or protocols developed are not hindered by cultural differences. For example, requiring an Asian nurse to confront a physician without first providing assertiveness training may lead to problems. To help overcome cultural barriers to communication, some organizations develop very specific language that individuals can use to communicate with each other. For example, if a nurse says, "I feel uncomfortable," this signals to the rest of the surgical team that he or she senses a problem and that all work should stop until the problem is addressed. By giving the nurse specific words to use and reinforcing the response to those words, cultural differences can be overcome.[15]

DEALING WITH DISRUPTIVE BEHAVIOR

Loud music is blaring, an individual in the corner is swearing because he just dropped something on his foot, and two people nearby are raising their voices at each other. While this scene could be happening in a bar or raucous sporting event, it is, unfortunately, occurring in an OR, and the patient has just been brought in.

As mentioned before, the OR is full of countless disruptions, and many of them cannot be controlled. However, the ones that can be controlled, including the verbal language, body language, and tone of voice of perioperative team members, can and should be managed.

Disruptive behavior is anything that upsets the smooth flow of a situation. It could be an argument, a string of verbal expletives, or an aggressive action, such as a shove or push. Organizations should not tolerate such disruptions in the OR. They can compromise patient safety, decrease staff satisfaction, and set a tone that is contrary to effective communication.

Organizations can discourage disruptive behavior in a variety of ways, including the following:

■ ***Encourage conflict resolution.*** The nature of human interaction produces conflict. Such conflicts could include performance issues or disagreements about patient care. Effective teams know how to deal with conflict and, if possible, resolve it. Training and education on conflict resolution is necessary for a perioperative team to survive. Perioperative staff should be taught to discuss conflicts in a respectful yet firm manner. Teaching the staff to use phrases such as "I perceive an issue here…" or "It seems to me that…" can be helpful. Staff members who disagree should strive to reach common ground and common goals and offer opinions in a constructive manner.[16] If consensus is not possible, it is acceptable to agree to disagree, politely.

■ ***Enforce a "zero-tolerance" policy.*** Despite an organization's efforts at establishing a culture of open communication and teamwork, some individuals may not be able to control their anger and funnel it into an effective communication technique. Individuals such as these may still exhibit aggressive or intimidating behavior. They may forbid questions, shut down collaboration, and ignore input. If an organization's leadership does not enforce a zero-tolerance policy for this type of behavior, then all the team training in the world will be for naught. If the perioperative staff feels that abusive and intimidating behavior will be overlooked, then the motivation for perioperative team behavior is greatly reduced. Communication between team members may be stifled, and the organization may regress. Talking about effective communication is not enough. Both physicians and staff must realize that effective communication is required by the organization and that intimidation and power struggles will not be tolerated. For those individuals struggling with how to communicate effectively, personal coaching or mentoring may be helpful. This training gives the individuals someone with whom they can work directly to improve their skills. In addition, uncontrolled disruptive behavior may be a symptom of an underlying issue, such as substance abuse. (See Appendix 2.)

Using Technology to Improve Communication

In addition to structured conversations and team training, organizations can improve communication in the OR by using technology. Systems such as infrared tracking, a nurse-call system, a wireless telephone system, and a Web-based electronic greaseboard can all help staff members communicate more effectively, identify where patients are in the system, ensure that critical information is not lost or forgotten, and diminish the amount of time spent waiting for certain processes to complete. While these technical options are not appropriate for every organization, this discussion introduces the concepts surrounding them for consideration.

Infrared Tracking

An infrared tracking system helps track patients throughout an organization. Using such a system, an organization can keep patients efficiently moving through the perioperative process and ensure that no patient is misplaced.

How It Works. On the day of a patient's surgery, the patient is assigned a tracking badge. It is the size of a wristwatch and can be placed on the patient's wrist next to his or her ID bracelet. A unique badge number is assigned to the patient. The tracking badge tracks the patient's location and the amount of time spent in each location by emitting infrared signals to sensors mounted in the ceiling throughout the perioperative area. Each time a patient moves to a new area or completes a specific process, such as checking in or completing surgery prep, the system displays that information and records the information in a database. At any given time, any staff member can see where a patient is and how long he or she has been there.[17] Some organizations use this technology with an electronic greaseboard.

Telephone and Call Systems

Perioperative nurses spend most of their day running–running to and from the OR, the preoperative area, the postoperative area, patient rooms, equipment and supply rooms, and so forth. Some organizations have used an in-house telephone and call system to help streamline communication among these nurses who are on the go.

How It Works. Each preoperative nurse is given a wireless telephone, which can be assigned to a specific set of rooms. A call-button system is used to text message the phone whenever the patient arrives in a new location or completes a process. For example, when the nurse-call button in the patient room is pressed, a text message can be sent to the telephone assigned to that room. Also, when the patient arrives in the preoperative area and the registrar completes the registration process, the registrar can push a call button to send a text message to the nurse, notifying him or her that the patient has been checked in. The call-button/phone text messaging system can be used at a variety of points throughout a patient's surgical experience, including when the patient is prepped for surgery, when the perioperative team is ready for surgery, when the room is ready for surgery, and so forth. In addition, OR housekeeping staff members can be given alphanumeric pagers that let them know when an OR must be cleaned. They can then use the nurse-call-button system to notify the nurse that the room has been cleaned.[17]

Electronic Greaseboard

An electronic greaseboard is like a chalkboard posted on an organization's intranet site. It allows information about patients to be added, subtracted, and supplemented in real-time, so that perioperative team members can quickly view a patient's status. An electronic greaseboard can integrate information from the previously described infrared tracking system and nurse-call system. Electronic medical records can also be integrated, if warranted. To view an electronic greaseboard, organizations may choose to use a Web-based application and display it over monitors strategically located throughout the perioperative environment.[17]

Ensuring Effective Handoff Communication

A handoff communication is a contemporaneous, interactive process of passing patient-specific information from one caregiver to another (or from one team of caregivers to another) for the purpose of ensuring the continuity and safety of the patient's care. Examples of handoff communication that occur in the OR include (but are not limited to) the following:

- Communication between staff at surgical shift changes
- Communication with other units, such as when clinicians transfer care to a postanesthesia care unit (PACU) or an intensive care unit (ICU)
- Communications between the OR and other departments, such as pathology or the laboratory

Traditionally, patient handoffs are a high-risk time in the OR when critical information can get lost, forgotten, or misplaced. Handoffs are a leading type of communication breakdown that can result in patient injuries.[18]

The primary objective of any patient handoff is the accurate transfer of information about a patient's status and care plan. Effective handoff communication includes the following characteristics:[19]

- Is face-to-face
- Is verbal and interactive
- Follows a standardized order, where communication is concise, efficient, and to the point
- Occurs with minimal interruptions
- When possible, is collaborative
- When necessary, involves the reading back of critical information
- Provides clinical and situational context
- Includes assessments of current situations and recommendations for future situations
- Incorporates relevant medical history, lab reports, vital signs, and so forth
- Is well documented

National Patient Safety Goal 2E. Implement a standardized approach to handoff communications, including an opportunity to ask and respond to questions.

An important way to ensure effective communication during patient handoffs is to standardize the handoff process, which is required by National Patient Safety Goal 2E (see above). There are many ways to standardize the handoff process, and not every approach will work for all organizations or all departments. Because handoff communications typically involve multiple disciplines, organizations should consider creating a multidisciplinary group to examine handoffs within the organization, identify areas of risk, and address those areas through standardization. Organizations should involve perioperative staff members, ask them what they consider important information to communicate in a handoff, and gain their input on the standardization process. Organizations should ensure that standardized handoff communication protocols are easy to follow and do not add to the work burden.

Meghan Dierks, M.D., suggests that the following components be included in surgery handoff communications:

- ***Baseline metrics and benchmarks.*** These would include, but not be limited to, the patient's behavioral, emotional, and mental status; medication needs and status; and prior perioperative experiences.
- ***Information on the most recent phase of care.*** This would include, but not be limited to, a description of the procedure; whether it went according to expectations, and if not, why not; whether there were any unexpected events; the implications of these events; and concerns about specific aspects of the patient's care.
- ***Current status.*** This would include at least the patient's vital signs, mental status, pain control, and airway breathing status, and whether these are within expectations. This section should also identify emergency trends and the anticipated course of vital signs, mental status, airway status, and so forth.
- ***Expectations for the next phase of care.*** This would include, but would not be limited to, expectations for the patient's status, progress, readiness for discharge, rate of change for vital signs, and patient goals.
- ***Other issues.*** This could include, but would not be limited to, medication dosages, patient room number, location of the family, and whom to contact.[18]

☞ **Tip:** Unlike some processes in the OR, handoff communications should not be done using a checklist because checklists can potentially limit the handoff conversation and leave out important context.[18] ☜

Situation-Background-Assessment-Recommendation

The Situation–Background–Assessment–Recommendation (SBAR) technique is an easy-to-remember, concrete mechanism that is useful for framing any conversation, especially critical ones, requiring immediate attention and action.[20] Organizations may want to consider incorporating the SBAR technique into their handoff communication processes. Developed by Michael Leonard, M.D., SBAR is used to standardize the type of information shared between team members and helps set the expectation that specific informational elements will be communicated every time a patient is discussed. This ensures that the relevant and important pieces of clinical information are communicated every time. SBAR stands for the following:

- ***Situation***–What is going on with the patient?
- ***Background***–What is the clinical background or context?
- ***Assessment***–What do you think the problem is?
- ***Recommendations***–What would you do to correct it?[21]

According to Dr. Leonard and colleagues, "The S-B-A-R model is particularly helpful in situations where a nurse–physician encounter must occur. It helps get both parties on the same page, as the physicians want to focus on the problem and the solution, and the nurses know they will be expected to relate aspects of the problem. S-B-A-R sets the expectation that critical thinking associated with defining the patient's problem and formulating a solution occur before the physician is contacted. Thus, both parties know that the conversation will include the assessment and recommendation for care that is relevant to the patient's current status."[3]

In addition to structuring handoff communications, the SBAR methodology can be helpful in structuring communication between other departments, such the laboratory, pharmacy, or pathology departments.

Communicating with Patients

A recent survey of nearly 7,000 adults ages 18 and older found that nearly 20% of respondents had at least one communication problem the last time they visited a physician. Respondents stated that they left the visit with questions they wanted to ask but did not; many felt the physician only partially listened or did not listen at all to what they said. Many respondents said they understood only a little of what the physician told them. One quarter of respondents indicated that they did not feel as involved in medical decision making as they would like to be, and nearly one third said they did not have as much time with their physicians as they desired.[22]

An organization may not need a national survey to reveal patient–staff communication problems. Patient satisfaction surveys, focus groups, and direct observation may all show that patients do not do feel involved in medical decision making and do not feel that the staff is communicating well.

Historically, communication between a patient having surgery and the surgical team has not been stellar. Patients often do not see their physicians before surgery, the nurse seems rushed to complete the preoperative assessment, and patient questions and concerns may not get addressed. In some cases, patients are so intimidated by the process that they fail to ask what they want to know or to communicate valuable information that the surgical team would benefit from knowing.

There are a variety of ways that organizations can encourage effective communication between the perioperative staff and surgical patients. The following sections discuss some of these strategies.

Encouraging Perioperative Dialogues with Patients

As mentioned in the introduction to this publication, perioperative care occurs before, during, and after the operative phases of a patient's surgery. Within all three of these phases, effective communication should occur between the surgical team and patient.

The most logical staff member to communicate with patients during the perioperative process is the nurse. Nursing visits during the perioperative period provide patients with opportunities to express expectations, share concerns, receive information, and discuss anxieties. These meetings can help relieve anxiety and stress and ensure that the patient and the perioperative team enter the operative process with the same information.[23]

To help ensure effective communication during the perioperative process, organizations may want to encourage the use of a perioperative dialogue. This is a model that is structured around three meetings with the patient and a designated nurse. The first meeting occurs when the patient and nurse meet before the surgery, at which time the patient is given the opportunity to get to know the nurse, tell his or her story, and describe his or her thoughts and expectations concerning anesthesia and surgery. During this time, the nurse listens, answers, and documents the patient's questions and/or concerns. This is a time for allaying patient fears and also for identifying issues that could come up in surgery. Both patient and nurse can learn and benefit from this exchange.

During the intraoperative period, the second meeting occurs. This happens when the patient meets the same nurse in the preoperative and operative or procedure area, and the nurse again explains what is going on and implements any planning he or she has done with the patient. At this time, the nurse may want to introduce the patient to the members of the perioperative team. Putting names with faces can make the perioperative team seem more human and assure the patient that he or she is indeed well cared for.

The third meeting occurs in the postoperative area. Again, the same nurse visits the patient. This time the dialogue focuses on the patient's experiences. This would be an appropriate time to discuss any unusual events that occurred, such as if the patient felt something during surgery (see Chapter 9).[23]

One of the most important aspects of the perioperative dialogue is familiarity. The fact that the same nurse meets with the patient before, during, and after surgery provides continuity for the patient and encourages the understanding that someone on the perioperative team cares about and is looking out for him or her. Nurses participating in a perioperative dialogue should actively listen to the patient and respond to his or her concerns. Using a story format to explain what is going to happen, what is happening, and what the future will hold can be particularly comforting to patients.[23]

If the nurse identifies information within the perioperative dialogue that could benefit the perioperative team, it is critical that he or she be empowered to share that information with team members. Surgeons, anesthesiologists, and other nurses can all benefit from this knowledge and should be willing to hear and respond to it.

Educating the Perioperative Staff on How to Effectively Communicate with Patients

To help ensure proper communication with patients, organizations should consider offering education for physicians, nurses, and other perioperative team members on the importance of including the patient in his or her care. Such education programs should also provide tips on how to communicate effectively with patients, including asking open-ended questions, observing and interpreting body language, encouraging the patient's questions and feedback, and using terminology that is easy for the patient to understand.

Physicians, in particular, need focused training on how to effectively communicate with patients. Medical schools do not typically teach listening skills or encourage physicians to empathize with their patients. Organizations may need to overcome the educational experiences of physicians to improve patient–physician dynamics. To elicit physician participation in patient communication training, an organization may have to emphasize the benefits of good patient–physician communication. Such communication may positively affect patient outcomes and reduce the risk of malpractice litigation, which is linked to poor physician communication.[24]

Using Appropriate Communication Tools

Unfortunately, many individuals in the United States cannot read above a fifth-grade level. Most health care information is written way above that level. Consequently, a gap exists between the information provided and the information understood. To help overcome this issue, organizations should design written materials that are easy to read and understand. Organizations should also offer non-written alternatives to illiterate individuals. When providing verbal education, the perioperative staff should use simple phrases and easily understood vocabulary. As previously mentioned, stories can be particularly helpful in communicating with patients about their surgery.

Effective patient–staff communication can mean the difference between a positive experience in a health care organization and a negative one. Organizations that carefully develop communication strategies that involve patients in their care, respond to patient concerns, and communicate information quickly and clearly can help make patient–staff interaction positive and effective.

Sidebar 1-5. Communicating with the Pediatric Patient

Although patients of all ages may have a variety of emotional responses to surgery and the OR, children may be particularly vulnerable to fear. Organization staff who work with pediatric surgical patients should be trained on how to communicate effectively with these patients. For example, staff members should refrain from talking to one another about a child as if he or she is not there. In addition, communication should be in positive and happy tones, and quiet voices are preferable to loud ones. Consider the following example:

> As a four-year-old requiring surgery is brought into the OR, the nurse and anesthesia care provider greet her with smiles. The room is quiet and warm. The anesthesiologist and nurse work to get the patient comfortable on the OR bed and ask her questions about how she is feeling. They ask if she is excited for the ice cream that is sure to follow the procedure. The scrub nurse and the surgeon greet the child warmly and present a balloon with a smiley face to the distracted, but nervous, child. The child is then asked to say her A-B-Cs while looking at the balloon and is told she will wake up in a little while. The child is as comfortable as she can be, given the circumstances.

References

1. Vincent C., et al.: Systems approaches to surgical quality and safety: From concept to measurement. *Ann Surg* 239: 475–482, Apr. 2004.

2. Joint Commission on Accreditation of Healthcare Organizations: *Root Causes of Sentinel Events.* http://www.jointcommission.org/NR/rdonlyres/FA465646-5F5F-4543-AC8F-E8AF6571E372/0/root_cause_se.jpg (accessed May 9, 2006).

3. Leonard M., Frankel A., Simmonds, T.: *Achieving Safe and Reliable Healthcare.* Chicago: Heath Administration Press, 2004.

4. Bianchi-Sand S.: It takes a team to prevent errors. *Am J Nurs* 103:89–90, Dec. 2003.

5. Sutcliffe K., Lewton E., Rosenthal M.: Communication failures: An insidious contributor to medical mishaps. *Acad Med* 79:186–194, Feb. 2004.

6. Makary M., et al.: Operating room briefings: Working on the same page. *Jt Comm J Qual Patient Saf* 32:351–355, Jun. 2006.

7. Air Force's 11 patient safety tools build better team communication. *OR Manager* 19:1, 12–13, Feb. 2003.

8. Mound T.: Closing the communication loop: Using read-back/hearback to support patient safety. *Jt Comm J Qual Saf* 30:460–464, Aug. 2004.

9. Consortium Management LLC & McNeary, Inc.: Chain of command. *Compass: Navigating Today's Healthcare Risks & Liabilities* 2:1–4, Mar. 2004.

10. Violato C., et al.: Multisource feedback: A method of assessing surgical practices. *BMJ* 326:546–548, Mar. 8, 2003.

11. Farbstein K.: Role-playing case simulations: A tool to improve communication and enhance safety. *Jt Comm J Qual Saf* 29:261, May 2003.

12. Wise H., et al.: *Making Health Teams Work.* Cambridge, MA: Ballinger Publishing Company, 1974.

13. Beyea S.: Learning from stories: A pathway to patient safety. *AORN J* Jan. 2004.

14. Spath P. (ed.): *Partnering with Patient to Reduce Medical Error.* Chicago: American Hospital Association, 2004.

15. Personal communication with Michael Leonard, M.D., director of Patient Safety for Colorado Permanent Medical Group, Colorado Springs, Colorado, Oct. 1, 2004.

16. Fenson S.: *Do As I Say: Quick Tips for Masterful Communication.* Ivy Sea Inc., Mar. 2000. http://www. pf.inc.com/articles/2000/03/18145.html (accessed Oct. 20, 2004).

17. Bahlman D.: Using technology to improve and support communication and workflow processes. *AORN* 82:65–73, Jul. 2005.

18. Help on handoffs: Meeting JCAHO's new 2006 patient safety requirement. *OR Safety* 21:1, 3, 8–12.

19. Patterson E.: Handoff strategies in settings with high consequences for failure: Lessons for health care operations. *Int J Qual Health Care* 16:125–132, Feb. 2004.

20. Kaiser Permanente of Colorado: *SBAR Technique for Communication: A Situational Briefing Model.* http://www.ihi.org/IHI/Topics/PatientSafety/SafetyGeneral/Tools/SBARTechniqueforCommunicationASituationalBriefingModel.htm (accessed Jun. 21, 2006).

21. Leonard M., et al.: The human factor: The critical importance of effective teamwork and communication in providing safe care. *Qual Saf Health Care* 13:185–190, Oct. 2004.

22. Rollins G.: Medical errors, poor communication undermine quality of care, patient satisfaction. *Report on Medical Guidelines & Outcomes Research* May 17, 2002, pp. 5–7.

23. Lindwall L.: Patients' and nurses' experiences of perioperative dialogues. *J Adv Nurs* 43:246–253, Mar. 2003.

24. Levinson W., et al.: Physician–patient communication: The relationship with malpractice claims among primary care physicians and surgeons. *JAMA* 277:553–557, 1997.

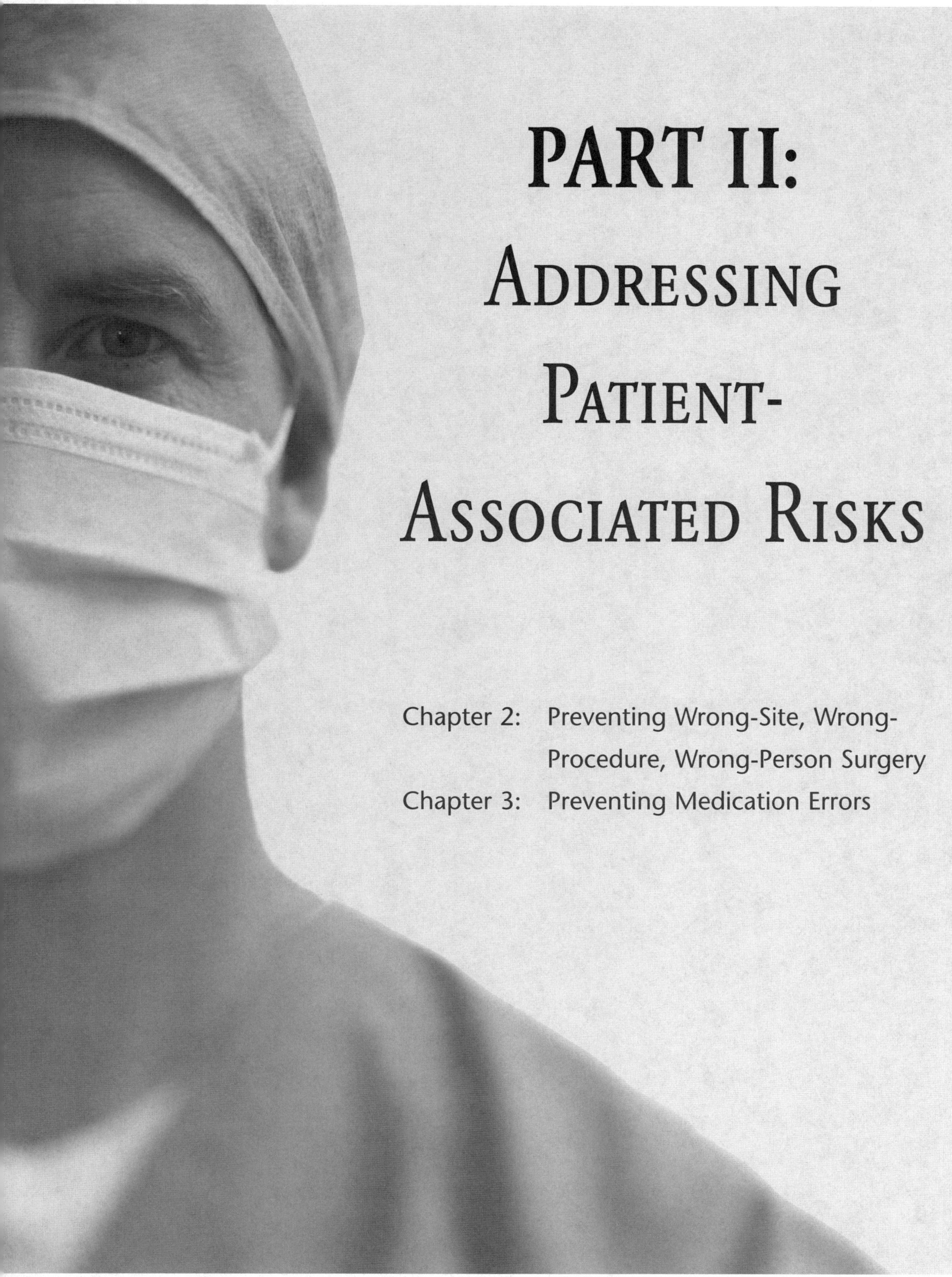

PART II: Addressing Patient-Associated Risks

Chapter 2: Preventing Wrong-Site, Wrong-Procedure, Wrong-Person Surgery

It Could Have Happened This Way

It was spring, and Max Whitney, * *a 17-year-old high school student, was supposed to be playing baseball on his school's varsity team. A candidate for a baseball scholarship, he was hoping to impress the college scouts. Unfortunately, Max was sitting on the bench, with his pitching arm in a sling, due to an error that could have been prevented. A few months prior, Max had fallen and hurt his left elbow. At the time, his physician applied a splint to his arm. Although the arm seemed healed, Max still experienced limited motion and a great deal of pain. His physician recommended surgery to repair the injury. The recovery time was expected to be a few weeks, well before the start of the baseball season. Since Max used his right arm for pitching, the surgery was not supposed to impact his ability to play baseball.*

Unfortunately, a mix-up in the operating room (OR) caused Max's hopes about baseball scholarships to fade. The surgeon was in a hurry to get started on the surgery, the nurse was out of the room when the surgeon marked the site, and Max's medical record was not readily available for surgical site confirmation. Instead of operating on his left arm, the surgeon operated on his pitching arm, which was, in fact, perfectly normal. Although Max was expected to recover, it would not be in the few weeks that were initially anticipated. He would not have time before the start of the baseball season to strengthen his pitching arm and was in fact recommended to sit out for the duration to let both arms completely heal.

Across town, the surgeon who operated on Max was devastated. He was well respected in the community and specialized in pediatric surgery. Hospital leadership was also reeling. No one thought an event like this could occur at a reputable health care organization with well-trained and well-intentioned staff.

Although Max's story is hypothetical, there are too many events like his that are all too real, and many of them have even more devastating consequences–as, for example, when a surgeon amputates the wrong leg, a mastectomy is performed on a patient who should have had a lumpectomy, or an invasive cardiac electrophysiology study is performed on the wrong patient.

Despite intensive educational efforts by medical, nursing, and health care leadership organizations, including the Joint Commission and the Association of periOperative Registered Nurses (AORN), wrong-site surgery events continue to occur. In the 12 years it has been collecting data on sentinel events, the Joint Commission has received, as of March 31, 2006, reports of 470 events involving wrong-site surgeries, making this type of sentinel event the second most frequently reported. What is even more disturbing is that these data may represent only a small portion of the actual total number of wrong-site surgeries because significant underreporting is likely.[1]

At the most fundamental level, every physician, surgical team member, and organization leader must recognize that failures involving wrong-site, wrong-procedure, and wrong-person surgery *could* happen anywhere. The American Academy of

* All patient and care-provider names in this book are fictitious.

Orthopedic Surgeons (AAOS) reports that during a 35-year career, an orthopedic surgeon has a one in four chance of performing a wrong-site surgery.[2] No one who works in the surgical arena is immune to the problem, and, therefore, everyone must maintain a high degree of awareness of the possibility of this type of event.

Why Wrong-Site, Wrong-Procedure, and Wrong-Person Surgery Occurs

Despite the fact that no one thinks these types of events will occur in one's own organization, there are several reasons they do. Ineffective or inadequate communication between members of the surgical team, lack of patient involvement in site marking, and lack of procedures for verifying the operative site can all increase the chances of a wrong-site surgery. Some other risk factors include the following:

- Inadequate patient assessment. This inadequacy could include having no formal procedure for verifying the surgical site; the absence of any oral communication in the verification procedure; or the lack of a final check of the correct procedure, site, and patient in the OR.
- Inadequate medical record review. This situation could occur because relevant information sources are not available in the OR or the surgical staff fails to review these information sources in the immediate preoperative period.
- Exclusion of certain surgical team members in the verification process
- A culture that does not support open communication between surgical team members and fosters the idea that the surgeon is always right and cannot be questioned.
- Problems related to illegible handwriting
- Use of abbreviations related to the surgical procedure, site, or laterality

While a wrong-site surgery event can occur during any kind of surgery, the chances increase during the following situations:

- More than one surgeon is involved in the case, either because multiple procedures are contemplated or because the care of the patient is transferred to another surgeon.
- Multiple procedures are conducted on the same patient during a single trip to the OR, especially when the procedures are on different sides of the patient.
- Unusual time pressures exist, related to an unusual start time or pressure to speed up the preoperative procedure, such as during an emergency situation.
- The patient has unusual characteristics, such as a physical deformity or massive obesity, that might alter the usual process for equipment setup or positioning of the patient.

In these types of situations, it is particularly important to have effective communication between surgical team members, including any surgeons working on a case. In addition, a comprehensive site marking and verification process that is consistently followed can help.

☞ **Tip:** Like other specialties, ophthalmology is not immune to surgical errors. Although it occurs rarely, one type of error is the insertion of an incorrect intraocular lens (IOL) in cataract surgery. The IOL can be the wrong power, wrong size, or wrong type, which can lead to postoperative refractive errors and less satisfactory vision for the patient. Some actions that can lead to incorrect IOL placement include the following:

- Use of an older, outdated IOL formula
- Incorrect measurement of the eye's axial length (biometry)
- Incorrect keratometry readings
- Mistakes in entering the data into the IOL calculation program
- Incorrect labeling or packaging of the IOL by the manufacturer
- Mistakes in providing the IOL at the time of surgery, such as a mix-up with an IOL for another patient

To help minimize the likelihood of incorrectly placed IOLs, organizations should consider the following actions:

- Ensure that all technicians working with IOLs are well trained in biometry and keratometry.
- Check calibration of equipment on a daily basis.
- Perform measurements for the IOLs in both eyes, for comparison. If at all possible, consider printing only one eye per IOL report to avoid confusion and the possibility of selecting the IOL not intended for the operative eye.
- Repeat or double-check all eye measurements, if appropriate.
- Have the IOL calculation report or form that contains patient information available in the OR so that the staff can double-check calculations and patient information.
- Invest in an automated software program that performs calculations, provides IOL selections, and documents patient visual outcome.
- Encourage and foster good communication among the surgeon/assistant surgeon and OR personnel.
- Have the correct lens in the OR prior to sedation/anesthesia.
- Perform a final verification of the IOL power and diopter against the IOL calculation report before incision or before IOL insertion.

For more information on preventing the incorrect placement of IOLs, see the American Academy of Ophthalmology Web site, at http://www.aao.org.

Source: American Academy of Ophthalmology: *Minimizing Wrong IOL Placement.* http://www.aao.org/education/library/safety/iol.cfm (accessed Jul. 16, 2006.)

Understanding the Universal Protocol

The fact that wrong-site, wrong-procedure, and wrong-person surgeries continue to happen is disturbing; however, the good news is that they can be prevented. To help organizations with this goal, the Joint Commission created the Universal Protocol for Preventing Wrong Site, Wrong Procedure, Wrong Person Surgery™. The Universal Protocol emphasizes the importance of communication before the start of surgery and is endorsed by 51 professional health care associations and organizations. Compliance with the Universal Protocol is required by all accredited hospitals, ambulatory care organizations, and office-based surgery facilities.

The Universal Protocol was created to reduce the likelihood of wrong-site, wrong-procedure, and wrong-person surgeries by providing a process to assist the surgical team in preventing mistakes. It applies to all operative and other invasive procedures that expose patients to more than minimal risk, including procedures done in settings other than the OR, such as a special procedures unit, an endoscopy unit, or interventional radiology suites. The protocol is made up of the following components:

- A preoperative verification process
- Marking of the surgical site
- A time-out that is held immediately before the start of a procedure

---

UP 1. The organization fulfills the expectations set forth in the Universal Protocol for Preventing Wrong Site, Wrong Procedure, Wrong Person Surgery and associated implementation guidelines.

- **Requirement 1A.** Conduct a preoperative verification process as described in the Universal Protocol.
- **Requirement 1B.** Mark the operative site as described in the Universal Protocol.
- **Requirement 1C.** Conduct a "time-out" immediately before starting the procedure as described in the Universal Protocol.

(For the complete Universal Protocol and its implementation guidelines, go to http://www.jointcommission.org/PatientSafety/UniversalProtocol/.)

The Preoperative Verification Process

The purpose of this process is to ensure that all relevant documents and studies are available prior

Prevent Wrong-Site Quality Management Sheet

HOSPITAL NAME					PATIENT LABEL
Prevent Wrong Site Quality Management Sheet					
BEFORE DAY OF SURGERY	DATE	*Items to Verify*	*Verified (yes or no)*	*Signature*	*Missing or Corrected Information*
Registrar		Physician Office Sheet, Advantx			
Admitting Interview		Physician Office Sheet, Advantx, Patient Interview, OR Schedule, OR Consent			
Preadmit/Preop RN		Patient Interview, OR Schedule, OR Consent and Diagnosis, History & Physical			
DAY OF SURGERY	DATE	*Items to Verify*	*Verified (yes or no)*	*Signature*	*Missing or Corrected Information*
Check In		Physician Office Sheet, Advantx, Patient Interview, OR Schedule, OR Consent			
Preop Surgical Nurse		OR Schedule, Patient Advantx Sheet, Patient Interview, OR Consent and Diagnosis, History & Physical			
Preop Holding Nurse		OR Schedule, Patient Advantx Sheet, Patient Interview, OR Consent and Diagnosis, History & Physical			
OR Circulator		OR Schedule, Patient Advantx Sheet, Patient Interview, OR Consent and Diagnosis, History & Physical, Patient Prepped Area, OR Setup			
RETURN COMPLETED QUALITY MANAGEMENT SHEET TO MANAGER					

Used with permission of Mary Shiels Hospital.

Figure 2-1. *This checklist, developed by staff at Mary Shiels Hospital, a USPI facility in Dallas helps staff verify paperwork at several steps throughout the patient process.*

Source: Joint Commission Resources: *Perspectives on Patient Safety,* Jun. 2005, p. 6, with permission from Mary Shiels Hospital.

to the start of a procedure, that they are reviewed, and that they are consistent with each other, the patient's expectations, and the team's understanding of the intended patient, procedure, site, and (as applicable) any implants. Any missing information or discrepancies must be addressed during this time.

The preoperative verification process should be an ongoing process of information gathering and verification, beginning with the determination to do the procedure, and continuing through all settings and interventions involved in the preoperative preparation of the patient.

To be effective, a preoperative verification process should verify the correct person, procedure, and site at the following times:

- At the time the surgery/procedure is scheduled
- At the time of admission or entry into the facility
- Any time the responsibility for care of the patient is transferred to another caregiver
- Before the patient leaves the preoperative area or enters the procedure/surgical room

☞ **Tip:** To help ensure a thorough preoperative verification process, organizations may want to consider creating a checklist that can be used to ensure availability and review of any relevant documentation, images, implants, or special equipment. Although such a checklist is not required by the Joint Commission, it is a common approach to ensuring a comprehensive, consistent, and effective preoperative verification process. Figure 2-1 is an example of such a checklist. ☜

Sidebar 2-1. Marking the Surgical Site

The Universal Protocol prohibits marking any non-operative site(s) unless necessary for some other aspect of care. This is because of incidents like the following example:

> A surgeon was going to repair a man's ruptured right Achilles tendon. The patient was anesthetized while he was lying on his back, and the left leg was marked "No." While the surgeon scrubbed, the scrub nurse and anesthesiologist turned the patient on his stomach. After scrubbing, the surgeon returned to the same side of the table, not realizing the patient had been flipped over, causing the leg to change positions. The surgeon made an incision on the wrong side before realizing his mistake.[3]

Marking the Surgical Site

Organizations must mark the surgical site to unambiguously identify the intended site of incision or insertions. The mark should be visible after the patient has been prepped and draped.

To effectively mark the site, the person performing the procedure should make the mark at or near the incision site (see Sidebar 2-1, above). When making the mark, staff members should use a marker that is sufficiently permanent to remain visible after completion of the skin preparation. Adhesive site markers should not be used as the sole means of marking the site.

In addition to making the mark sufficiently permanent, the mark must also be unambiguous. For example, organizations may want to use the surgeon's initials or "YES" to mark the site. When differentiating between sides of the body, organizations should consider writing out "LEFT" or "RIGHT." It is important to note that an "x" could be considered ambiguous and should be avoided. Regardless of the method chosen for marking, the method and type of mark should be consistent throughout the organization.

Sidebar 2-2. Who Marks the Surgical Site

While ideal, it is not always possible for the person performing a procedure to mark the surgical site. In these cases, another member of the surgical team who is fully informed about the patient and the intended procedure must do the marking. Any delegation of responsibility for marking the surgical site must be consistent with applicable law and regulation.

While not all surgical sites will need to be marked, organizations, at a minimum, should mark all cases involving laterality, multiple structures (for example, fingers, toes, lesions), or multiple levels (for example, the spine). For a specific list of those sites exempt from surgical site marking, go to http://www.jointcommission.org/NR/rdonlyres/DEC4A816-ED52-4C04-AF8C-FEBA74A732EA/0/up_guidelines.pdf. If a patient requires multiple surgeries while in a health care organization, the site mark should be removed at the end of the procedure, unless the next surgical procedure will continue on the same site.

Involving the Patient in Marking. When possible, the marking process should involve the patient. To be done in a meaningful way, this procedure should happen before the patient is significantly sedated. Participation may be precluded by

Sidebar 2-3. The Sterility of Surgical Site Marking

There has been some concern that marking the surgical site could introduce bacteria, thus affecting the sterility of the site. A recent study showed "the prevalence of laboratory-detected micro bacterial growth to be 0% in cultured specimens obtained from prepared marked and unmarked surgical sites."[4] Like other studies, this confirms that there is no apparent infection risk with surgical site marking.

a disease state or heavy sedation; however, sedation by itself does not necessarily prevent the patient from participating in the site-marking process. The patient's capacity to participate must be based on an individual assessment. In cases of nonspeaking, comatose, or incompetent patients, or children, the patient involvement in the site-marking process should be handled in the same way as the informed-consent process. Whoever has authority to provide informed consent for the patient to undergo the procedure would, as appropriate, participate in the site-marking process.

While rare, the patient always has the right to refuse site marking. This situation should be handled in the same way as any other refusal by a patient offered care, treatment, or services. The organization's responsibility is to provide the patient with information to understand why site marking is appropriate and desirable and thus grasp the implications of refusing the site marking. If a patient refuses site marking, the organization should have a procedure in place to appropriately identify the site. This may involve a bedside consult with the surgical team, so everyone understands the location of the site and the patient's reluctance to be marked.[5]

☞ **TIP:** To help ensure surgical site marking, organizations should ensure the availability and use of appropriate surgical marking pens. Several manufacturers offer special surgical marking pens and other direct skin-marking methods that will not be washed off when the surgical site is prepped but can be removed following the procedure. For patients with dark skin, a marker other than black or dark blue should be used to ensure that the site marking is clearly visible. Some organizations have found red markers to be helpful.[6] ☜

> **Sidebar 2-4. Clarifying the Universal Protocol**
>
> Certain routine minor procedures, such as venipuncture, placement of a peripheral intravenous line, insertion of a nasogastric tube, and insertion of a Foley catheter, are not within the scope of the Universal Protocol and its requirements.

THE TIME-OUT

The purpose of the time-out, or "huddle," immediately before starting a procedure is to conduct a final verification of the correct patient, procedure, site, patient position, and, as applicable, availability of implants and special equipment. The time-out requires active communication between team members in a fail-safe mode–that is, communication is interactive, and the procedure is not started until any and all questions or concerns are resolved.

The time-out must involve the entire surgical team, including, at a minimum, the surgeon, anesthesia provider, and circulating nurse. Participation by the other members of the team, as appropriate to their involvement in the procedure, is also encouraged. In particular, there should be no barrier to anyone speaking up if there is concern about a possible error. To include some members of the team but not others sends the wrong message. It could imply that the input and opinions of some of the staff are not valued. During the time-out process, if a situation arises in which there is a discrepancy in staff response, an organization should have processes and systems in place for reconciling those differences.

A time-out must be conducted in the location where the procedure will be done, just before starting the procedure. It also must be briefly documented, such as in a checklist. The time-out must, at least, include verification of the following:

- Correct patient identity
- Correct side and site
- Agreement on the procedure to be performed
- Correct patient position
- Availability of correct implants and any special equipment or special requirements

Verification, site marking, and time-out procedures should be as consistent as possible throughout an organization, including in the OR and other locations where invasive procedures are performed.

☞ **Tip:** In the previous paragraphs, the word *must* is used in some places, and *should* is used in others. This is intended to distinguish between requirements (those steps that must be done as stated in the Universal Protocol) and preferred approaches that should be followed whenever possible, but each organization should allow some degree of flexibility to accommodate the unique characteristics of a particular organization. ☜

Strategies for Compliance with the Universal Protocol

Health care organizations can comply with the Universal Protocol by using a variety of strategies and processes; however, merely putting procedures in place will not ensure compliance with the protocol. Organizations must work with the staff, especially physicians, to develop appropriate and feasible processes, create a team atmosphere, provide effective training, and monitor compliance.

Using a Collaborative Approach to Development

If a process is cumbersome, confusing, or overly repetitive, organization staff may not follow it. Conversely, if a process is quick, easy, and straightforward, staff members will more likely follow it, particularly if they see the benefits of doing so. Developing processes for identifying and verifying the surgical site should be a multidisciplinary process, including representatives from the anesthesia, medical, and nursing staff. This collaborative effort not only helps ensure a feasible process but also reduces the amount of resistance to ongoing, consistent, and correct implementation of the policy.[7]

As leaders of the perioperative team, surgeons, in particular, are important to include in the development of thorough and time-efficient verification and marking processes. Physicians know what site-marking practices are prevalent in the physician community. Some practices could be effective, and others would need to be addressed through education efforts. Learning physicians' thoughts about ways to perform the processes without increasing time pressures and delays is critical. Physician involvement need not be limited to one particular hospital or organization. Physician leaders can encourage consensus on implementation across organizations throughout a particular community.

Educating Staff

Education and training on wrong-site surgery and the processes required for its prevention must involve all staff members. Each member of the surgical team has a unique role to play in the prevention of wrong-site, wrong-procedure, wrong-person surgery, and each member should be well aware of that role and how to play it. The staff should be familiar with the risk factors that could lead to wrong-site surgery, the prevalence of the problem, and the protocols and procedures in place to prevent it.

While wrong-site, wrong-procedure, and wrong-person surgeries occur more often than they should, they are not a daily occurrence in an organization. To make the threat of such an event more real, organizations should consider discussing actual occurrences to help drive home the issue. Organizations may want to share examples of near misses, offer statistics about the number of wrong-site surgeries occurring in the United States, and share media reports of actual cases and discuss how they occurred.

The staff should also be familiar with the components of the Universal Protocol and how those components are implemented within the organization. For example, if a checklist is used in the organization, the staff should be familiar with how to use that checklist and where it is found. Conference calls, regular in-services, role-playing exercises, posters, buttons, and tent cards can all be helpful in building awareness of and training staff on the organization's implementation of the Universal Protocol.

MAKING IT AN ORGANIZATIONWIDE ISSUE

Although surgical team members must be aware of their organization's interpretation of the Universal Protocol, they are not the only staff members who should be familiar with the topic of wrong-site surgery. Every staff member can play a role in preventing this type of error. Misinformation or errors in patients' paperwork, such as medical records, physician orders, lab reports, and so forth, can potentially lead to wrong-site errors. It is the responsibility of all organization staff to review and cross-check patient paperwork. When incorrect or missing information is discovered, it should be corrected immediately, and all individuals involved with the patient should be made aware of those corrections. To encourage the staff to look for errors, an organization might consider hosting a contest to identify paperwork mistakes.

Any staff member should be able and encouraged to verify information with a patient. The best way to do this is through open-ended questions. For example, instead of asking a patient, "Mr. Smith, we are operating on your left arm today, right?" a safer approach would be "Mr. Smith, what are we doing today?" or "Mr. Smith, what is your understanding of what Dr. Jones is doing today?"

EDUCATING PATIENTS

Staff members are not the only individuals who must be educated on the Universal Protocol. Patients who are having surgery and their families must understand what staff are doing to prevent wrong-site surgery and why they are doing it. Not only does this encourage patient involvement, but it can also prevent noncompliance with the protocol. Consider this example:

> An organization was having trouble with its physicians not marking the surgical site. In reviewing data on these cases, the organization found that the protocol was not being followed because the marking ink was upsetting to patients, especially those with breast cancer. The organization now takes into account the emotional issues of a patient's illness, especially breast cancer, and engages in discussions with patients before marking ink is used. Patients are educated on and made part of the process, and they understand that it is critical to their safety.[8]

Monitoring Compliance with the Universal Protocol

With any process or procedure, the only way to know if it is being followed is to monitor compliance. There are a variety of ways to monitor compliance with the Universal Protocol, including regular observation, medical record review, checklist review, and so forth. One organization chose to monitor compliance with its time-out procedure by surveying staff members as to whether the surgical team actually implemented the time-out before surgery and whether there was any reluctance or backlash from other members of the surgical team during implementation.[7]

When monitoring compliance, some data to collect could include the following:

- Number or percentage of preoperative checklists completed
- Number or percentage of sites unmarked
- Number or percentage of time-outs completed

Just collecting data is not sufficient. Organizations should regularly review compliance data and determine whether staff members are complying with the protocol, and if they are not, why not. Organizations that discover compliance issues may want to do a failure mode and effects analysis (FMEA) or a root cause analysis (RCA) to determine the underlying reasons behind noncompliance.

References

1. Joint Commission Resources: Physicians and the Universal Protocol. *Joint Commission: The Source* 2:3–4, Sep. 2004.

2. American Academy of Orthopedic Surgeons: *Sign Your Site: Wrong-Site Surgery.* http://www5.aaos.org/wrong/viewscrp.cfm (accessed May 10, 2006.)

3. Adapted from An example in wrong-site protocol is more than one year old, but problems mount. *Same Day Surg* 29: 121–132, Nov. 2005.

4. Cronen G. et al.: Sterility of surgical site marking. *J Bone Joint Surg Am* 87:2193–2195, Oct. 2005.

5. Effective strategies to comply with Joint Commission's Universal Protocol. *Hosp Peer Rev* 28:149–150, Nov. 2003.

6. American Health Consultants: Effective strategies to comply with Joint Commission's Universal Protocol. *Hosp Peer Rev* 28:149–150, Nov. 2003.

7. Schiedt R.: Home study program: Ensuring correct site surgery. *AORN J* 76:769–782, Nov. 2002.

8. Mawji Z., et al.: First do no harm: Integrating patient safety and quality improvement. *Jt Comm J Qual Improv* 28:373–386, Jul. 2002.

Chapter 3: Preventing Medication Errors

It Could Have Happened This Way

Jane Montgomery was undergoing surgery to treat her recently diagnosed facial cancer. During a portion of the surgery, the surgeon reached for an unmarked bottle on the sterile field. Certain that he knew the contents of the vial, the surgeon drew a sample of the fluid and injected it into Ms. Montgomery's spine. Not only was the vial not labeled, no one on the perioperative team verified the contents of the bottle. Everyone accepted the word of the surgeon as to the contents. Unfortunately, the fluid in the bottle was glutaraldehyde. The injection caused Ms. Montgomery to sink into a coma, leaving her brain-dead.[1]

Medication error in surgery, like the one that occurred in the above example, can have devastating consequences. One study shows that medication errors in surgical settings are more likely to lead to patient harm than those that take place outside surgical settings.[2] Surgery takes place in a fast-paced environment in which complex care is provided using high-tech equipment and high-alert drugs. These factors come together to set the stage for medication errors. Recognizing the seriousness of medication errors and the potential for them to occur in an environment that is full of distractions, organizations should implement safe medication practices in the operating room (OR).

It is important to recognize that medication management in the OR is very different from that in a typical inpatient clinical unit. In the OR, medications and solutions must be delivered aseptically to the sterile field, and product packaging may prevent delivering a medication aseptically in its original container. Inconsistent and illegible labels or a lack of any labels can contribute to medication errors.

Potential system issues in the OR, such as inconsistent practices, unclear written policies, and lack of training or education, can also increase the risk of patients receiving an unintended medication or solution.

Safe Medication Practices in the OR

To help organizations develop policies that support safe medication use in the OR, the Association of periOperative Registered Nurses (AORN) developed guidelines based on the five "rights" of medication use[3]:

1. Right patient
2. Right medication
3. Right dose
4. Right time
5. Right route

These guidelines strongly encourage organizations to develop a protocol of standardized procedures that support safe medication practices. This section discusses some of those safe medication practices.

The Joint Commission's Medication Management (MM) standards and National Patient Safety Goals 2 and 3 also require organizations to implement safe medication practices. For reference,

* All patient and care-provider names in this book are fictitious.

some of the relevant standards and goals are included throughout this chapter.

Delivering Medications Safely

To ensure the safe delivery of medications to the OR, staff members should handle medications in an aseptic manner, which means not using intermediate containers to transfer medications from nonsterile to sterile fields. In addition, the perioperative staff should not remove stoppers from vials in order to pour the medications. Medications should be dispensed directly from the original container to a syringe, using a sterile vial spike, filter straw, or plastic catheter. A single medication should be delivered, prepared, and labeled before a second one is prepared.

Labeling Medications on and off the Sterile Field

One of the most important aspects of the MM process in the OR is the proper labeling of medications. All medications and medication containers on the sterile field should be labeled, including the container that contains the medication or solution and the device used to deliver the solution. Errors have resulted from medications and other solutions being removed from their original containers and placed into unlabeled containers, as medications or other solutions in unlabeled containers cannot be identified.

To help highlight the issue of proper medication labeling, the Joint Commission created National Patient Safety Goal 3D, which requires organizations to properly label all medications, medication containers (for example, syringes, medicine cups, basins), or other solutions on and off the sterile field in perioperative and other procedural settings.

Even if only one medication is used in a surgery, it should be properly labeled. Although this precaution may seem unnecessary, if additional medications are added to the field during a procedure, an error could occur if the original medication has not been properly labeled. In addition, confusion and miscommunication can occur when the scrub person is relieved during a procedure. The surgeon could pick up an unlabeled syringe and administer the wrong medication.

National Patient Safety Goal 3D. Label all medications, medication containers (for example, syringes, medicine cups, basins), or other solutions on and off the sterile field.
MM.4.30. Medications are labeled.
MM.6.20. The organization responds to actual or potential adverse drug events and medication errors.

There are several ways organizations can properly label medications and medication containers. For example, organizations can use commercially printed or handwritten medication labels or write directly on a medication container with an indelible marker. Sterile markers used for labeling medication containers on the sterile field can be purchased from a medical manufacturer and should be in sterile packages that allow aseptic transfer to the sterile field. It is not recommended that a health care facility sterilize an ordinary office product marker for the purpose of labeling medications on the sterile field.[4]

According to the implementation expectations of the National Patient Safety Goals, medication labels should include the drug name, strength, amount (if not apparent from the container), expiration date when not used within 24 hours, and expiration time when expiration occurs in less than 24 hours.

☞ **Tip:** Different kinds of labels are appropriate for various types of syringes. For example, sterile medication labels may adhere best to asepto syringes because there is more flat surface on the cylinder of the syringe on which to apply the label. Labels may not adhere well to bulb syringes because the surface of such syringes is round, flexible, and often has ridges. Labels that do not adhere adequately to a bulb syringe could increase the risks of leaving

Sidebar 3-1. What Should Be Labeled?

The following medications and solutions fall under National Patient Safety Goal 3D and should be labeled when removed from their original packaging:

- Prescription medications
- Sample medications
- Over-the-counter drugs, herbal remedies, dietary supplements, and vitamins
- Nutriceuticals
- Vaccines
- Diagnostic and contrast agents used on or administered to persons to diagnose, treat, or prevent disease or other abnormal conditions
- Radioactive medications
- Respiratory therapy treatments
- Parenteral nutrition
- Blood derivatives
- IV solutions (plain or with electrolytes and/or drugs)
- Chemicals and reagents such as formaline, saline, sterile water, Lugol's solution, radiopaque dyes, glutaraldehyde, and chlorhexidine
- Chemotherapeutic medications
- Any other product designated by the Food and Drug Administration (FDA) as a drug

foreign bodies in the surgical wound (see Chapter 10). Consequently, using a sterile indelible marker to write directly on the bulb may be a more suitable strategy for labeling bulb syringes.[4] ☜

☞ **Tip:** It is important to keep all original medication containers as well as delivery devices in the OR for reference purposes until a procedure has been concluded. After the procedure is complete, the contents held in all labeled containers on the sterile field must be discarded properly. In addition, current and reliable medication reference materials, including full package inserts, calculation charts, and updated physician preference cards, should be made readily available to practitioners in the OR. ☜

☞ **Tip:** Medication manufacturers can play an important role in OR medication safety. For example, organizations should work with manufactures to use distinctive labels for medications and solutions. In addition, manufacturers should be encouraged to provide medications and solutions in ready-to-use or ready-to-administer containers that allow for sterile delivery of medications to the sterile field.[3] ☜

☞ **Tip:** All medications in the OR should be labeled at all times. Any medications lacking an identification label should be discarded. ☜

Verifying Medication Labels

Verbally and visually verifying medications are effective ways to ensure that the right medication is getting to the patient. Within this verification process, each medication's name, strength, dosage, and expiration time, when applicable, should be verified. Verification is completed by two qualified individuals before the start of a procedure. If only a circulating nurse is present, he or she should be empowered to confirm the medication with another licensed professional (for example, the anesthesiologist or surgeon) performing the surgical procedure.

Standard MM.4.40. Medications are dispensed safely.

Shift changes can be particularly risk prone for medication errors because staff members are entering and exiting the OR, and communication can break down. To prevent confusion at shift change or break relief, entering and exiting personnel should note and verify concurrently all medications as well as their labels on and off the sterile field. Communication about the amount of medication administered should take place at this time as well.[3]

IDENTIFYING THE PATIENT AND CONFIRMING ANY MEDICATION ALLERGIES

Before a medication is administered, the perioperative staff should verify the patient's identity using two identifiers (as per National Patient Safety Goal 1A). This can be done through the patient's identification band, if appropriate. If the band is not visible due to positioning or draping, the circulating nurse and anesthesia care provider may verify the patient's identity by using the patient's record.

Standard MM.1.10. Patient-specific information is readily accessible to those involved in the MM system.
National Patient Safety Goal 1A. Use at least two patient identifiers (neither to be the patient's room number) whenever administering medications or blood products, taking blood samples and other specimens for clinical testing, or providing any other treatments or procedures.

During patient verification, the circulating nurse should also verify the presence of any allergies. The staff should check the patient's chart for allergies and reconfirm any allergies with the patient before surgery. Verifying allergy information can be challenging in the OR, as the pace of a procedure is quick and often there is no time to do an extensive review of the medical record to determine allergies. Organizations may want to consider conducting a proactive risk assessment on this aspect of the medication use process to determine what problems exist with obtaining, documenting, and quickly retrieving allergy information.[2] Organizations that provide a specific location in the medical record for the documentation of allergy information can help ensure that such information can be easily and quickly located before surgery. Electronic medical records are particularly helpful in this process, as organizations can designate a particular field for allergy information.

☞ **TIP:** To easily communicate a patient's allergies, some organizations choose to use allergy bands. These bands offer a clear warning that a patient has allergies and can serve as a reliable source of information for clinicians, even when the patient is medicated, sedated, or confused. To be effective, allergy bands must be readily available for use throughout an organization in a standardized manner. If the use of bands is haphazard, the staff will not be able to reliably differentiate when a patient does not have allergies versus when a staff person forgot to put on the band. Policies regarding allergy bands should clearly identify who assesses allergies, when the allergy band is applied, and how the allergy band is updated and verified. In addition, the information on the band should be readily visible, pertinent, and readable.[5] ☜

☞ **TIP:** During a procedure, the patient's response to an administered medication should be continually monitored for evidence of an adverse reaction. ☜

☞ **TIP:** To ensure patient safety, organizations should treat chemicals and reagents, such as formalin, saline, radiopaque dyes, and glutaraldehyde, in the same standardized way as medications. As can be seen in the example at the beginning of this chapter, a mistake with these substances can be just as harmful as a medication error. ☜

Strategies to Reduce the Risk of Medication Errors

While the OR can be ripe for medication errors, these are not a foregone conclusion. That said, organizations should not simply ask perioperative team members to "be more careful." Medication safety requires a system approach that minimizes the risks and helps perioperative team members provide consistent and appropriate care. Some system strategies organizations can use to prevent medication errors are discussed in the following sections.

ESTABLISHING FORCING FUNCTIONS

To minimize risks related to medication administration, organizations should consider developing constraints or forcing functions that make it difficult to make a medication error. An example of this would be a dose-limit protocol that sets limits on doses of particular medications. If a dose exceeded the limit, it would be questioned. Special permission would be required to exceed a dose limit. Other forcing functions include automatic stop orders or removing certain medications, such as cytotoxic agents, from the OR.

Educating the Staff on Safe Medication Practices

Creating policies and procedures that embody safe medication practices is important. However, if organization staff members are not aware of the need for such policies and do not understand how to thoroughly follow these policies and procedures, they will be ineffective. Without appropriate training, staff members can unintentionally contribute to errors. To keep patients safe, perioperative staff should be up to date on medication safety issues and how to effectively address those issues.

As with other kinds of issues, there are a variety of ways to train the perioperative staff on medication safety issues, policies, and procedures. As previously mentioned, organizations that use a variety of methods for training staff can help everyone learn effectively and efficiently. For example, simulations and role-playing exercises are very effective teaching tools because staff members must participate in a simulated situation and perform their responsibilities as if it were a real-life issue. Actually performing a procedure may help the staff better understand and learn it.

To provide background information on medication errors and the importance of medication safety, organizations may want to offer information via e-mail or on Web sites for those staff members who are comfortable with technology. Other effective methods for sharing information include in-services, posters, newsletters, and videotapes. Sharing statistics, stories, and scenarios may drive home the point that following policies and procedures to ensure medication safety may help save lives.

Education is important, and organizations must make sure that staff members understand what they have learned and fulfill their roles accordingly. As previously mentioned, one way to do this is by administering tests to the staff after training activities. Tests can be written, verbal, or electronic. They can involve multiple-choice questions or require staff members to demonstrate their knowledge. Whatever the design of the test, its purpose is to verify that staff understand the training and can perform their roles and responsibilities. Tests can also be used to identify areas for further staff education. Other ways to verify staff knowledge include checklists, weekly evaluations, and observations.

If testing reveals a lack of understanding or a lack of compliance, it is important that the organization revisit its training efforts in that area. For example, if observation reveals that the perioperative team is not verifying the labels on medications before surgery, the organization should revisit its training efforts in this area. This may involve changing the content of the training. It could also involve changing the way training is conducted.

Creating a Reporting System

The only way to get a true picture of the number and nature of medical errors occurring in the OR is to institute a reporting system that captures that information. Reporting systems can capture all kinds of errors, including medication errors. They should not be designed to blame people but to identify system vulnerabilities that can lead to lapses in patient safety. Reporting systems can take many forms but should all have the following characteristics:

- ***Are nonpunitive.*** Staff members must feel comfortable reporting and be assured that their reports will not lead to retribution, disciplinary action, or other negative consequences. If staff members feel they will be blamed for the issues they report, then they will not report them.

- ***Are easy to use.*** If staff members find it difficult or time-consuming to report errors, then they will be less likely to report them. The staff must know what to report, who to contact, how to contact them, and when to contact them. Organizations should clearly define what constitutes a reportable medication event and share that definition with the staff.
- ***Use multiple venues.*** Organizations that use multiple venues for reporting, such as e-mail, telephone hotlines, focus groups, surveys, personal meetings, or written reports, also yield higher reporting rates.
- ***Examine more than just errors.*** By encouraging staff members to report both errors and near misses, organizations can learn what errors occurred as well as what errors could have occurred. In addition, the staff should report conditions that cause concern because those conditions could lead to near misses or errors.
- ***Contain feedback mechanisms.*** Staff members who report should receive feedback about their report, including an acknowledgement of the report, what was learned about the problem, what is being done to address the problem, and the results of improvement efforts. If an organization does not respond to reports, then staff members may question the need for a reporting system. Staff members must feel that by reporting, they are helping to improve the safety of patients and to decrease medical errors in the OR.

Although the data collection aspect of reporting systems is important, analyzing those data is even more critical. Organizations should take information reported within a reporting system, examine that information, look for patterns and trends, and address any issues that arise. Appropriate statistical techniques for this analysis could include run charts, control charts, histograms, Pareto charts, fishbone diagrams, and so forth. Trending error data over time can be helpful to show improvement or areas that need further work.

☞ **TIP:** Comparing medication error rates to those of other organizations has limited benefits because all organizations should be working to eliminate medication errors and not just be better than their neighboring hospitals. ☜

AVOIDING POTENTIALLY DANGEROUS ABBREVIATIONS

As previously mentioned, surgery takes place in a fast-paced environment where critical and sometimes life-saving information must be communicated quickly and clearly. Unfortunately, when health care providers use some abbreviations, the communication can be quick but may be far from clear.

The issue of dangerous abbreviations is by no means new. The Institute for Safe Medication Practices (ISMP) has been discussing this topic for more than 25 years. The Joint Commission addresses the issue of dangerous abbreviations in both the standards and the National Patient Safety Goals. In the second element of performance (EP) of Standard IM.3.10, it states that abbreviations, acronyms, and symbols should be standardized throughout an organization, and a do-not-use list should be created and distributed to all the staff. The do-not-use list is also a requirement of the second National Patient Safety Goal.

National Patient Safety Goal 2B. Standardize a list of abbreviations, acronyms, and symbols that are not to be used throughout the organization.

Standard IM.3.10, EP 3. Minimum data sets, terminology, definitions, classifications, vocabulary, and nomenclature, including abbreviations, acronyms, and symbols, are standardized throughout the organization.

To create a standardized do-not-use list, organizations should consult with the do-not-use list of abbreviations, acronyms, and symbols from the Joint Commission. (See Sidebar 3-2 on page 39.) While these are the only abbreviations, acronyms,

Sidebar 3-2. Joint Commission Do-Not-Use List*

Do Not Use	Potential Problem	Use Instead
U (unit)	Mistaken for "0" (zero), the number "4" (four) or "cc"	Write "unit"
IU (International Unit)	Mistaken for IV (intravenous) or the number 10 (ten)	Write "International Unit"
Q.D., QD, q.d., qd (daily) Q.O.D., QOD, q.o.d, qod (every other day)	Mistaken for each other Period after the Q mistaken for "I" and the "O" mistaken for "I"	Write "daily" Write "every other day"
Trailing zero (X.0 mg)** Lack of leading zero (.X mg)	Decimal point is missed	Write X mg Write 0.X mg
MS MSO_4 and $MgSO_4$	Can mean morphine sulfate or magnesium sulfate Confused for one another	Write "morphine sulfate" Write "magnesium sulfate"

* Applies to all orders and all medication-related documentation that is handwritten (including free-text computer entry) or on preprinted forms.

** Exception: A trailing zero may be used only where required to demonstrate the level of precision of the value being reported, such as for laboratory results, imaging studies that report sizes of lesions, or catheter/tube sizes. It may not be used in medication orders or other medication-related documentation.

and symbols that organizations are required to enforce the staff not to use, organizations can certainly add to this list. It is important to note that organizations should exercise caution in adding new elements to the do-not-use list if staff members in their organization practice at multiple organizations. Different do-not-use lists could cause confusion.

Once an organization has identified its dangerous abbreviations, it is critical that the staff be educated on the do-not-use list. While there are a variety of ways organizations can do this, following are some suggestions that have proved useful in some health care organizations:

- Print the list on brightly colored paper, sticky notes, posters, stickers, or magnets and place it in medical records, at/on/near computers, and/or in patient care areas.
- Provide pocket-sized cards displaying the list to staff members.
- Print the list in the margin or at the bottom of the physician order sheets and/or progress notes.
- Delete prohibited abbreviations from preprinted order sheets and other forms.
- Create clipboard covers that provide the list.
- Provide the list on the front page of the intranet.
- Provide a card with the list that can be attached to the back of the identification badge.
- Place tent cards with the list where physicians write orders and dictate.
- Have the list printed on pens.
- Send monthly reminders of the list to the staff via e-mail.
- Create an educational display for use during Patient Safety Awareness Week.
- Place articles in organization newsletters to the staff.
- Provide mouse pads with the list.
- Conduct a mock survey and question staff members to test their knowledge.
- Work with the organization's software vendor to ensure that changes are made to be consistent with the list.

- Create a catchy name or theme, such as Do the "Write" Thing; Dirty Dozen; Outlaw Abbreviations–Join the Patient Safety Posse; "Operation BANEM" (Banned Items); or a Smokey the Bear–style poster saying "You can prevent a fatal error."
- Promote a "Do-not-use abbreviation of the month" campaign.
- Create a slide show/presentation illustrating poor handwriting and dangerous abbreviations. Include actual examples from the organization.

STORING MEDICATIONS

Securely storing medications so as to ensure their stability and safety is critical to the provision of safe and quality care. If medications are not stored properly, they may become unstable and present a health hazard to both patients and staff. In addition, proper storage of medications can help prevent any inadvertent mix-ups with look-alike and sound-alike medications. Also, by limiting multiple concentrations of medications in a facility, organizations can help prevent patients from receiving the wrong concentration of medicine.

Standard MM.2.20. Medications are properly and safely stored.

Appropriate storage of medication involves preserving the medication's chemical stability, documenting its expiration date, noting any damage or contamination, and including any appropriate warning labels. Organizations should have procedures in place to quickly recall a medication if necessary or segregate a medication that has expired. Including the medication lot number and expiration date on the packaging can help with these efforts.

☞ TIP: To help keep track of expiring medications, some organizations tag items that are going to expire in the next 60 to 90 days with a brightly colored circle or stamp. This can alert staff members to select those products first and use them only if they will be administered before the expiration date. ☜

Segregating Sound-Alike/Look-Alike Drugs. With tens of thousands of brand-name and generic drugs currently on the market, the potential for error as a result of confusing drug names is significant. One way to reduce the risks associated with sound-alike/look-alike drugs is to appropriately segregate them when storing them. The Joint Commission is committed to this issue and has made the segregation of sound-alike/look-alike drugs a part of its MM standards and also a requirement of National Patient Safety Goal 3.

National Patient Safety Goal 3C. Identify and, at a minimum, annually review a list of sound-alike/look-alike drugs used in the organization and take action to prevent errors involving the interchange of these drugs.

The Joint Commission requires organizations to identify and annually review a list of sound-alike/look-alike drugs used in the organization and take action to prevent errors involving the interchange of these drugs. Actions may include segregating the medications through separate boxes, different shelves, different cabinets, or the door versus a shelf in a refrigerator or freezer. A list of look-alike/sound-alike drugs can be found at http://www.jointcommission.org/NR/rdonlyres/C92AAB3F-A9BD-431C-8628-11DD2D1D53CC/0/lasa.pdf. Organizations should go through that list and define at least 10 pairs of drugs available in the organization that must be addressed.

Removing Concentrated Electrolytes. Although many organizations are aware of the risks to patient safety related to concentrated electrolytes, patient deaths continue to occur. To prevent these egregious circumstances, organizations must remove all concentrated electrolytes from patient care units unless the organization can show that the presence of one of these products is an absolute requirement for patient care, such as in

ORs where cardiac surgery is performed. If concentrated electrolytes are present, special safeguards must be taken to prevent their inadvertent administration. These safeguards include limiting access to the medications, placing special warning labels on the products, segregating them from other medications that are directly administered, and establishing and maintaining a par level system.

Inspecting Medication Storage Areas. All areas where medication is stored should be periodically inspected. These include refrigerators, freezers, crash carts, anesthesia carts, trauma kits, and floor stocks. While the Joint Commission does not dictate the frequency of the inspection, organizations should define the frequency and content of the inspection in a written policy and comply with that policy.

Some things to look for in an inspection include appropriate temperature, lighting, humidity, and ventilation. The cleanliness of the area should be observed, as should the level of security. In addition, organizations should check for the proper segregation of drugs and other chemicals.

Using a Verbal Order Policy

Verbal orders are often used in the surgical setting and present significant risks, such as the administration of inappropriate dosages or incorrect medications. Verbal orders in the OR are subject to misinterpretation for a variety of reasons, including masks worn in the OR that muffle speech or music playing that can impede the clear receipt of a verbal order. In addition, the person receiving a verbal order may be distracted due to the complex and hectic nature of the surgical environment.

To help prevent medication errors, all verbal orders in the OR should be verified out loud and repeated back. For example, a surgeon could say "Give Mr. Smith 10,000 units of heparin" and the anesthesiologist should then repeat "10,000 units of heparin." The surgeon should then confirm that this is correct. Ideally, the circulating nurse or anesthesia provider should also write down a verbal order, when possible. For further verification, the individual administering the drug could announce the drug and dose just before administration.[6]

To further ensure that verbal orders are clear, organizations should encourage the staff to spell out words, using such aids as "*D* as in dog," or "*F* as in Frank." Staff should also use a digit-by-digit technique. This technique involves stating each digit of a number, such as saying "one-six" instead of "sixteen." The staff should also avoid abbreviations where possible. An organization should develop a verbal order policy for the OR, educate the staff on that policy, and monitor its consistent use.

National Patient Safety Goal 2A. For verbal or telephone orders or for telephonic reporting of critical test results, verify the complete order or test result by having the person receiving the order or test result read back the complete order or test result.

Identifying and Addressing the Use of High-Alert Medications

High-alert medications are drugs that bear a heightened risk of causing significant patient harm when they are used in error. Although mistakes may or may not be more common with these drugs than with others, the consequences of an error with these medications can be more devastating to patients.[7] Some examples of high-alert medications used in surgical settings include heparin, epinephrine, and phenylephine. Many of the adverse events associated with these drugs are related to incorrect dilutions or inappropriate concentrations.[8] However, negative events can result when medications are transferred to unlabeled vials or when indications and administration procedures are misinterpreted.

To reduce the risks associated with high-alert medications, organizations can take a variety of actions. In addition to following the safe medication practices discussed previously, organizations should also consider the following:

■ Standardize preparations of all diluted solutions. These standardized solutions should be prepared before surgery so they are in ready-to-use or ready-to-administer containers that allow for sterile delivery of a container's contents.

■ Refrain from diluting medications on the surgical field. Although organizations should refrain from using nonstandardized dilutions, a necessary dilution should be performed only by a registered pharmacist. If a medication must be diluted in the OR, diluting instructions should be precalculated by a pharmacist and readily available for clinical use.

■ Use only premixed solutions and limit the number of medication concentrations.

■ Have clinicians in the OR communicate verbally all doses of all medications to be administered and clarify the maximum dose with the anesthesia care provider and surgeon.

■ Educate nurses and unlicensed personnel about the safe handling and administration of high-alert medications.

■ Assess and validate clinical competency with handling and administering these medications.

RECONCILING MEDICATIONS

To "reconcile" is to compare and reach agreement. In the context of medications, reconciliation is the process of comparing the medications that the patient has been taking prior to the time of admission or entry to a new setting with the medications that the organization is about to provide. It is up to each organization to determine how this process takes place. Whenever and however the comparison takes place, it should take place early enough to improve the safety of the organization's MM processes and, hence, patient safety.

Whenever an individual moves from surgery into recovery, a complete and current list of that individual's medications–as obtained on admission and updated during surgery–should be communicated to the next provider of service. Medication reconciliation is important because medications have become more numerous and complicated, and incompatibilities between different drugs can cause serious issues.

When organizations do not effectively reconcile medications, patients may receive duplicate medications, incompatible drugs, wrong dosages, or wrong dosage forms. Reconciling medications can reduce these errors as well as the duplication of work because one person gathers the information and passes that information along to other caregivers rather than each caregiver gathering the same information. It can also ensure that the patient is receiving all medications necessary to his or her care and eliminate any medications that are no longer needed.

To address the issue of medication reconciliation, the Joint Commission created National Patient Safety Goal 8. The second requirement of this goal requires the perioperative staff to communicate a complete list of patient medications to the next provider of service, using a standardized process. All information communicated should be included in the patient's medical record.

National Patient Safety Goal 8B. A complete list of the patient's medication is communicated to the next provider of service when a patient is referred or transferred to another setting, service, practitioner, or level of care within or outside the organization.

To completely reconcile medications, staff members must compare what the patient is taking at the time of admission or entry with what the OR is planning to prescribe. Ideally, the staff in the OR will not prescribe a medication before attaining the complete list of medications that the patient is currently taking. However, if there is an urgent situation in which the resulting delay would harm the patient, perioperative staff can prescribe without attaining the complete list of current medications. After staff members stabilize the patient, they should take steps to compile a complete list of the patient's medications and compare it to the medications currently being provided.

Following are some suggestions on how perioperative staff and organization leadership can improve the medication-reconciliation process:

- Put the list of medications in a highly visible place in the patient's chart and include essential information about dosages, drugs schedules, immunizations, and drug allergies.
- Include the reasons for changing medications in the reconciliation information so that the next provider of care understands why the patient was transferred to a new type of medication.
- Include medication-reconciliation information in the change-of-shift procedures as well as in the physician's progress notes.
- Educate the community (possibly through the primary care physician) so that patients know to bring their medication lists as well as their insurance cards when entering a health care organization.
- Before automating a medication-reconciliation process, organization leaders must make sure there is a stable paper process. Leaders should also consider a careful design that designates who can enter medication information initially and who can update and change information so that new errors are not introduced.
- Keep the staff focused on reconciliation by producing a monthly report that honors staff members who caught potentially dangerous medication problems by reconciling.

For more information on medication reconciliation, organizations can find helpful resources at the Web sites of the Massachusetts Coalition (http://www.macoalition.org) and the Institute for Healthcare Improvement (http://www.ihi.org).

References

1. Based on Around the nation: Operating room error puts patient, 64, in coma. *The New York Times.* Mar. 6, 1985. http://query.nytimes.com/gst/fullpage.html?res=9E02E7DD1139F935A35750C0A963948260 (accessed Apr. 28, 2006).

2. Medication errors in the perioperative environment. *USP Patient Safety CAPSLink.* United States Pharmacopeia. Mar. 2003.

3. AORN guidance statement: Safe medication practices in perioperative practice settings. *AORN J* 79:674–676, Mar. 2004.

4. Connor R.: Clinical issues: Medication labeling. *AORN J* Nov. 2005. http://www.aorn.org/journal/2005/novci.htm (accessed May 1, 2006).

5. Beyea S.: Oops—The patient is allergic to that medication: Patient safety first. *AORN J* Mar. 2003. http://www.findarticles.com/p/articles/mi_m0FSL/is_3_77/ai_99237614/print (accessed May 1, 2006.)

6. Cohen M.: Verbal orders in the O.R.: Masking the Truth? *Nursing* 32:32, Jun. 2002.

7. Institute for Safe Medication Practices: *ISMP's List of High-Alert Medications.* http://www.ismp.org/Tools/highalertmedications.pdf (accessed May 1, 2006).

8. Beyea S.: Patient safety first alert: Epinephrine and Phenylephrine in surgical settings. *AORN J* 77:743–747, Apr. 2003.

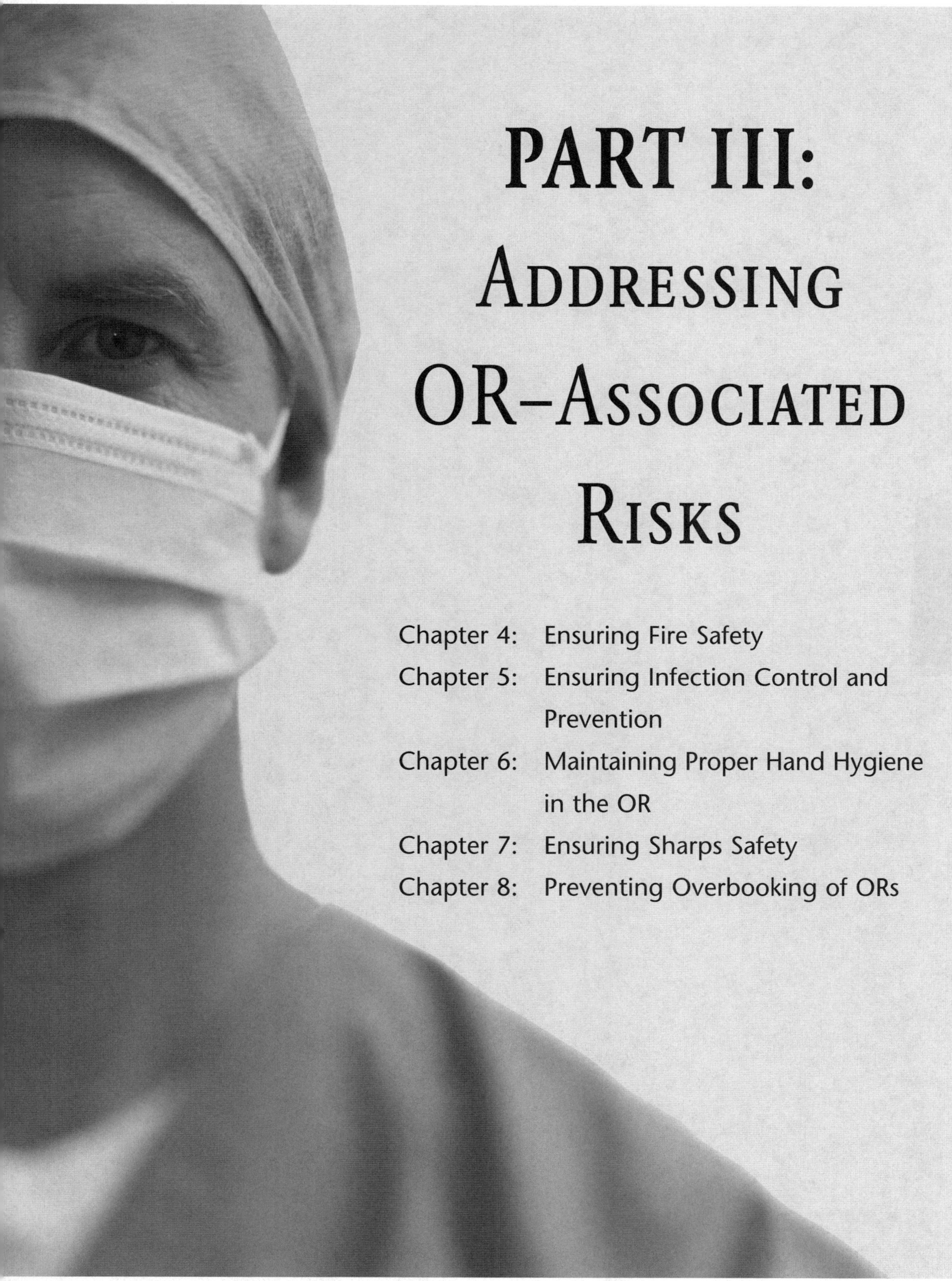

PART III: Addressing OR–Associated Risks

Chapter 4: Ensuring Fire Safety

It Could Have Happened This Way

Jessica Crumb needed an emergency C-section to safely deliver her baby. After being rushed into the operating room (OR), Mrs. Crumb was given epidural anesthesia. A nurse prepared the skin of her abdomen with an alcohol-based preparation. The nurse used a sponge to apply the colorless liquid and draped the area. Mrs. Crumb had some bleeding, and the surgical team used an electrosurgical unit for coagulation. As soon as the active electrode from the electrosurgery unit was activated, the staff heard a loud noise, smelled smoke, and Mrs. Crumb began complaining of heat. The anesthesiologist also felt the heat and saw a flame. The surgical team reacted quickly and put out the fire with water. Fortunately, the team was able to complete the C-section, and Mrs. Crumb gave birth to a healthy baby boy.*

In the postanesthesia care unit, clinical staff members noted significant burns to Mrs. Crumb's inner thighs and legs. She was treated for these burns in the hospital's burn unit and released. After several weeks, Mrs. Crumb made a full recovery; however, she was not able to fully participate in her son's homecoming. She also had regular nightmares about the event and sought psychological counseling.

For its part, the hospital conducted an extensive investigation and discovered that the fire was caused by a pooling of alcohol-based skin preparation fluid underneath the surgical drapes. This pooling produced a vapor, which was ignited by a spark from the active electrode tip. Because the nurse used a sponge for applying the skin preparation and that preparation was clear, a larger volume of skin preparation was used than was necessary. Fortunately, the drapes used on Mrs. Crumb were somewhat fire retardant, so a more significantly dangerous situation was prevented. In addition, the use of epidural anesthesia versus general anesthesia was a benefit.[1]

* All patient and care-provider names in this book are fictitious.

While surgical fires like the one that harmed Mrs. Crumb are rare, they do occur. In fact, each year there are approximately 100 surgical fires reported in the United States.[2] However, according to ECRI, an independent nonprofit health services research agency, this statistic most likely reflects a severe underreporting of the actual number of fires in the wide range of surgical settings.

Statistics show that the most common ignition sources of surgical fires are electrosurgery and lasers, and the most common locations of surgical fires are in the airway and on the head or face.[3]

The Fire Triangle

Although surgical fires do not happen often, they can have severe consequences, including injury to patients and staff as well as significant damage to surgical equipment.

To call attention to the issue of surgical fires, the Joint Commission released a *Sentinel Event Alert* in June 2003 detailing the root causes of surgical fires and providing suggestions to reduce the risk of occurrence. The *Sentinel Event Alert* focused on the concept of the "fire triangle." The three "sides" of the fire triangle are a fuel source, an oxygen source, and an ignition source. These three elements must interact for a fire to start. For example, if a patient is having facial surgery and receiving supplementary oxygen (oxygen source) at the same time that an active electrode (ignition source) is being used, the heat from the active electrode tip could ignite the oxygen. If this explosion touches the patient's drape (fuel source), a fire could occur.

The OR is ripe with components of the fire triangle. Following are some common examples of fuel, ignition, and oxygen sources found in the OR:

- ***Fuel sources.*** A fuel source is anything that feeds a fire and could include both obvious and less-than-obvious sources:
  - Alcohol-based prepping agents
  - Linens, such as drapes, towels, gowns, hoods, and masks
  - Dressings
  - Ointments
  - Surgical equipment and supplies
  - Topical anesthesia sprays
  - Povidone-iodine solutions
  - Synthetic sutures
  - Human tissue
  - Gastrointestinal or bladder gases
  - Tracheal tubes
  - Breathing circuits
  - Operating table mattresses
  - Body hair
  - Blood pressure cuffs
- ***Ignition sources.*** These typically give off heat and must get hot enough to ignite any fuel material:
  - Electrosurgical units (ESUs) or electrocautery units (ECUs)
  - Fiber-optic light sources and cables
  - Lasers
  - Defibrillators
  - Halogen lights
  - Drills
  - Heated probes
- ***Oxygen sources.*** Many surgical procedures are performed in what is labeled an "oxygen-enriched environment," which the National Fire Protection Association (NFPA) defines as an environment having an atmospheric oxygen concentration greater than 23.5%.[4] The higher the concentration of oxygen, the more serious a potential fire is likely to be. Medical-grade oxygen is the most obvious source of additional oxygen in the OR; however, nitrous oxide can also cause combustion and can result in fires as serious as those resulting from 100% oxygen.[5] Other oxidizers include medically compressed air and ambient air.

Sidebar 4-1. National Patient Safety Goal 11

In addition to releasing *Sentinel Event Alert* 29, the Joint Commission added the prevention of surgical fires to its National Patient Safety Goals. The goal requires ambulatory care facilities and office-based surgery practices to reduce the risk of surgical fires by educating staff members, including operating licensed independent practitioners and anesthesia providers, on how to control heat sources and manage fuels. In addition, the goal requires organizations to establish guidelines to minimize oxygen concentration under surgical drapes. Although the Joint Commission only requires compliance with this goal in ambulatory care facilities and office-based surgery practices, hospitals can also benefit from educating their staff about surgical fires and putting guidelines in place to minimize oxygen concentration.

National Patient Safety Goal 11. Reduce the risk of surgical fires.

Requirement 11A. Educate staff, including operating licensed independent practitioners and anesthesia providers on how to control heat sources and manage fuels and establish guidelines to minimize oxygen concentration under drapes.

☞ **Tip:** Many resources can help organizations identify what they can do to help prevent surgical fires. Some of these resources include the following:

- National Fire Protection Association (NFPA), http://www.nfpa.org
- ECRI, http://www.ecri.org
- American Society for Healthcare Engineering (ASHE), http://www.ashe.org
- American College of Surgeons (ACS), http://www.facs.org
- Association of periOperative Registered Nurses (AORN), http://www.aorn.org ☜

Minimizing the Risks Associated with the Fire Triangle

The components of the fire triangle are always present to some degree in a surgical setting. While this fact may seem unsettling, most surgical fires can be prevented if staff members understand how fires occur and take the proper precautions to avoid risky situations. The following sections offer some prevention strategies that organizations can use to help eliminate the risks of surgical fires and enhance patient safety.

Preparing Patients Appropriately

How a patient is prepared for surgery can have a significant impact on whether he or she is at risk for experiencing a surgical fire. When preparing a patient for surgery, all areas of the patient's body that could become fuel sources during a fire should be appropriately prepped. Some steps perioperative staff can take to appropriately prepare a patient for surgery include the following:

- Coat all exposed facial hair located near the surgical site, including any beards or mustaches, in a water-soluble jelly to make them nonflammable.
- For some facial surgeries, cover the patient's eyes with swabs soaked in sodium chloride.
- Be aware that alcohol-based skin preps are flammable and avoid pooling or wicking these flammable liquid preps.
- Let the prepped area dry completely before covering the site with a nonflammable drape. It is important to note that pooled or wicked liquid takes longer to dry.
- Caution patients against wearing perfume, cologne, or hairspray on the day of surgery, as these typically contain flammable components and represent a potential fire hazard. Written preoperative patient instructions can help reinforce this request.

Sidebar 4-2. ASHE's Perspective on Alcohol-Based Surgical Prep Solutions

A fire during surgery in Omaha, Nebraska, which severely burned a patient, prompted the Nebraska Fire Marshal and the Director of the Nebraska Department of Health and Human Services (HHS) to issue a ruling banning alcohol-based solutions in preparation for surgery when cautery or electrosurgery is considered.[2]

This ban, initially supported by the NFPA and enforced by the Centers for Medicare & Medicaid Services (CMS), led to concerns that fire marshals in other cities and HHS directors in other states would follow suit.[2] The Joint Commission, ECRI, and ASHE responded that alcohol is considered the "gold standard" for preventing infections following surgery but emphasized the need to educate health care workers about preventing fires in any OR.

ASHE argued that there are several essential elements involved in the care of surgical patients, and each of these elements can contribute to fires in the OR if not managed properly. Therefore, ASHE argued that managing the risk of each element, rather than removing any single element, is essential to preventing surgical fires.[6] ASHE proposed amended code language to NFPA 99 to include strict precautions for the use of alcohol-based surgical solutions. The NFPA voted to accept ASHE's proposed amendments, thereby agreeing that alcohol-based solutions can be used, as long as these strict safety precautions are followed to prevent surgical fires.[7]

Using Equipment Safely

In some surgical procedures, such as laser eye surgery and plastic surgery, surgical equipment can serve as an ignition source to a potential fire. To prevent fires during such procedures, organizations should make sure that the surgical team uses any equipment properly and "engages" the equipment only when working on the patient. For example, a laser should be placed in standby mode when it is not in use. Likewise, an ESU should be activated only when the active electrode tip is in view, and it should be deactivated before the active electrode tip leaves the surgical site. In addition, the active electrode tip should be kept in a holster when not in use. The active electrode tip should be kept clean to minimize the likelihood of sparking or burning of tissue debris.[5]

For fiber-optic light sources, ECRI recommends completing all cable connections before activating the light source and placing the equipment on standby mode when disconnecting cables.[8] It is important to note that an active fiber-optic cable should never be allowed to rest on a drape.

Some organizations have foot pedals that surgeons use to turn equipment on and off. These foot pedals should be phased out because it is too easy for the staff to accidentally "trip" the foot pedal and unintentionally engage the instrument.

☞ **Tip:** If a surgical team is using open oxygen (O_2) sources, the team should consider using bipolar electrosurgery when possible and appropriate, such as for cauterization during head and neck surgery. Bipolar electrosurgery creates less sparking or arcing than monopolar electrosurgery. In addition, members of the surgical team should never use insulating sleeves cut from catheters to cover electrosurgical active electrode tips, and they should never use electrosurgery to enter the trachea during tracheostomy. ☜

Maintaining Equipment

Faulty equipment can wreak havoc during surgery. For example, equipment that shorts out or sparks during surgery can cause a fire. All electrical equipment that is used in the OR should be calibrated and inspected frequently. In addition, an organization should have a maintenance plan for all equipment and document regular and preventive maintenance to ensure that equipment is safe to operate. Before beginning surgery, the perioperative staff should ensure that any electrocsurgical equipment or other equipment that could act as an ignition source has had the appropriate maintenance. Easy verification can take place through stickers placed on the equipment. In addition, perioperative staff should inspect all electrical cords for fraying before plugging them into outlets or using them on a patient.

☞ **Tip:** When choosing new equipment for the OR, organizations should consider its fire safety record and weigh any cost savings against potential safety issues. ☜

☞ **Tip:** Organizations should carefully monitor the amount of potentially explosive chemicals that are brought into the OR. For example, 10% buffered formalin is explosive, and organizations should consider prohibiting large volumes of it in the OR.[9] For more information on hazardous materials in the OR, see Appendix 1. ☜

Controlling Excess Oxygen

Surgical settings have higher-than-average concentrations of oxygen. Nearly 75% of flash fires are related to oxygen-rich atmospheres, and these types of fires occur only in environments with 50% oxygen concentration and above.[8] High oxygen concentrations are particularly hazardous in head, neck, and upper chest surgery because the heat source is typically near where oxygen is flooding the patient's face.[10]

Not all surgeries require 100% oxygen. In fact, organizations are encouraged to use the lowest possible concentration of inspired oxygen that ensures adequate oxygen saturation. As a general policy, organizations should consider using air or nitrous oxide at 30% or less for open delivery. Reducing the oxygen concentration reduces the risk for fire because the oxygen side of the triangle is reduced. By reducing the number of surgeries conducted in oxygen-rich environments, an organization can reduce the chance of a surgical fire.

☞ **TIP:** One way to reduce the amount of oxygen used in a procedure is to use pulse oximetry to monitor a patient and administer oxygen only when necessary.[11] ☜

To further control excess oxygen, surgical drapes should be tented during surgery so that oxygen cannot accumulate. Oxygen circuit connections should also be checked to make sure they are leak free. If possible, surgical staff should use a properly applied incise drape to help isolate head and neck incisions from O_2-enriched atmospheres and from flammable vapors beneath the drapes. Proper application of an incise drape ensures that there are no gas communication channels from the under-drape space to the surgical site.

☞ **TIP:** The staff should be cautious when open-mask oxygen is administered, as there is an increased fire risk due to the possibility of an oxygen-enriched atmosphere around the operative field. If possible, surgical staff members should stop supplemental O_2 at least a minute before beginning the use of electrosurgery, electrocautery, or laser surgery on the head or neck. ☜

ENGAGING STAFF IN THE FIRE-PREVENTION PROCESS

Each member of the surgical team–surgeon, anesthesiologist, and nurse–should play a role in preventing surgical fires. Staff members must be aware of their roles and know how to play them.

It may be helpful to assign each team member a side of the fire triangle that he or she is responsible for managing. For example, the nurse may be responsible for limiting and properly addressing fuel sources; the surgeon may be responsible for ensuring that surgical equipment is used properly, thus limiting ignition possibilities; and the anesthesiologist may be responsible for limiting the presence of excess oxygen. (See Figure 4-1.)

Sidebar 4-3. Oropharyngeal Surgery

To prevent oxidation during oropharyngeal surgery, perioperative staff members should use suction as near as possible to any potential breathing gas leak to scavenge the gases from the oropharynx of an intubated patient. Staff members should wet any gauze or sponges used with uncuffed tracheal tubes to minimize leakage of gases into the oropharynx, and they should keep these materials wet. Surgical team members should keep all moistened sponges, gauze, pledgets, and their strings moist throughout the procedure to render them ignition resistant.

Responding to a Fire

While fire prevention is critical to safety in the OR, organization staff should be prepared to fight a fire and treat a patient if a fire does occur. It is important to note that fire response in the OR is very different from the response that occurs in other areas of a health care organization. "RACE"–rescue, alarm, contain, extinguish–is the acronym that describes the fire response in which all health care staff have been drilled, and this approach makes sense in other fire situations. For example, a fire that starts in a supply closet would involve staff members first rescuing patients, sounding the fire alarm, containing the fire, and then extinguishing it, if possible. But in the case of an OR fire (unless it is so out of control that evacuation is necessary), the staff should do the following:

- Extinguish the fire by disconnecting the patient's oxygen and ventilating the patient with air.
- Contain the fire by removing any burning materials from the patient.
- Rescue the patient by caring for him or her.

Sidebar 4-4. Preventing Surgical Fires During Anesthesia Induction

When organizations use a desiccated absorbent, the exothermic reaction of the absorbent with carbon dioxide (CO_2) in the respiratory circuit or "canister" of an anesthesia machine can produce extreme heat that sometimes leads to actual fire. Although rare, at least eight cases have occurred in the United States, and additional cases have been reported overseas.[12] As a result, health care providers should heed this potentially dangerous situation by making sure they understand the potential risks, take steps to prevent anesthesia-related fires, and know what to do when extreme heat or fire is produced.

So far, the reported cases have occurred when patients were breathing through a circle system with an inline CO_2 absorbent that was apparently desiccated.[12] This is what happens: The interaction of desiccated absorbents with anesthetics produces temperatures of several hundred degrees centigrade. Sometimes, the increase can result in fire in the anesthetic circuit—which, in turn, can lead to serious patient injury.[13] The impulse to dismiss the potential risk is strong because absorbents and anesthetics have been used together for a long time—and breakdown reactions always occur but usually without causing an actual fire. However, milder reactions might occur and potentially harm patients as well. In some cases, patients absorb flammable and presumably toxic gases, causing pulmonary injury.[14] Certain conditions are especially likely to lead to such danger. Consider the following examples:

- Sevoflurane presents greater high temperature and fire risk than desflurane or isoflurane, according to a recent study.[15] Dangerously high temperatures and fire are far more likely to occur when desiccated absorbents are combined with sevoflurane than when combined with desflurane or isoflurane.[15] With desflurane or isoflurane, the increases in temperature are much smaller, are not sustained, and are consistent with the lesser amounts of degradation.
- Different anesthetic agents display different levels of vulnerability to degradation, making some more susceptible to fire. In a demonstration study, desflurane underwent the least degradation, isoflurane far more, and sevoflurane the most. Temperatures increased most rapidly and peaked earlier with desflurane, in part because the larger concentrations used drove the degradation reactions to completion earlier.[15]
- Anesthesiology machines that are left inactive for extended periods are more susceptible to fire. Many fire incidents have occurred on Monday morning, after the machines have been left inactive for extended periods. A low fresh gas flow rate over an extended period of nonuse may contribute to unexpected desiccation of CO_2 absorbent materials on the anesthesia machine.

According to a "Dear Healthcare Professional" letter from Abbott, anesthesia providers should consider the following measures, which might reduce the occurrence of adverse events:

- Replace the CO_2 absorbent if it might be desiccated because it has not been used for an extended period.
- Completely shut off the anesthesia machine at the end of clinical use or after any case when a subsequent extended period of non use is anticipated.
- Turn off all vaporizers when not in use.
- Verify the integrity of the packaging of new CO_2 absorbents before use.
- Periodically monitor the temperature of the CO_2 absorbent canisters.
- Monitor the correlation between the sevoflurane vaporizer setting and the inspired sevoflurane concentration. An unusually delayed rise or unexpected decline of Ultane concentration compared to the vaporizer setting may be associated with excessive heating of the CO_2 absorbent canister.
- Avoid degradation by replacing desiccated absorbent with fresh absorbent or by pouring water onto the desiccated absorbent.

For more safety precautions regarding anesthesia use, see Chapter 9 and Appendix 1.

This difference between situations demonstrates why it is so important to have a fire plan geared to the OR.

☞ **Tip:** When extinguishing burning materials, it is important to refrain from stomping on them because the materials could stick to shoe covers and ignite surgical scrubs.[16] The staff should drop burning materials to the floor and extinguish them with water or a fire extinguisher. ☜

Sidebar 4-5. Responding to Different Types of OR Fires

Although a fire in the OR may occur in a variety of places, the following are suggested responses to a surgical wound fire and a patient airway fire:

Surgical Wound Fire

- Shut down medical gases.
- Pour saline into the surgical site.
- Remove the surgical drapes to the floor, along with any material that may have been burning.
- Search for and extinguish any additional flames.
- If there is smoke in the room, determine whether it is necessary to evacuate.
- Save all materials for later investigation.

Patient Airway Fire

- Disconnect the breathing circuit from the tracheal tube.
- Remove the tracheal tube and have another surgical team member extinguish it.
- Remove any other segments of burned tube, such as any cuff protective devices that may remain in the area.
- Reestablish the airway and resume ventilating with air until certain nothing is left burning in the airway.
- Examine airway for extent of damage.
- Save all materials for later investigation.

Creating a Fire Plan

In creating a fire plan, one size does not fit all, and the plan to deal with surgical fires must be specific to the OR and differ from the fire plan for the rest of the organization.

To be effective, an organization's OR fire plan should assign specific duties to each member of the perioperative team, including the primary surgeon, the anesthesiologist, the scrub nurse, and the circulating nurse. The fire plan should establish what everyone on the team should do in extinguishing the fire and assuring the safety of the patient. The fire plan should also assume that the patient is in the middle of a surgical procedure when the fire breaks out and therefore is at extreme risk. The plan should determine exactly how the perioperative team will manage the fire quickly and address the interrupted procedure.

For an example of a fire plan, see Figure 4-1 on page 54.

☞ **Tip:** It can be helpful to keep a bowl of water or saline in the OR. In some cases, this may be enough to extinguish a small fire. If staff members douse part of the surgical drapes, all drapes covering the patient should be removed and the patient should be redraped. This ensures that no other sparks have occurred where they cannot be seen.[11] ☜

One component of a fire plan is the evacuation check sheet. This is a tool that can help account for all personnel in the OR and determine any medical equipment and supplies necessary for an evacuation. An example of such a check sheet can be found in Figure 4-2 on page 56.

☞ **Tip:** Organizations may want to have instructions on how to care for a burn patient in the OR. This could include instructions regarding fluid control, pain control, and wound control. Staff members can refer to these as they prepare the patient for transport to a burn unit or another area of the health care organization.[11] ☜

Responsibilities of the Surgical Team During a Fire Emergency

Within the OR

Primary surgeon

- Quickly remove any burning material from the patient.
- Work to extinguish the fire.
- Control any bleeding.
- Complete the surgical procedure as quickly as possible.
- Prepare to help evacuate the patient, if necessary.

Anesthesiologist

- Stop the flow of anesthetic gases if the fire is located on the patient or drapes.
- Shut off oxygen that is flowing directly to the patient.
- Ventilate the room with air.
- Assist with obtaining portable oxygen, suction machines, and other equipment, if necessary.
- Determine with the authority having jurisdiction whether the emergency gas shutoff values should be turned off.

Scrub nurse and circulating nurse

- Remain with, protect, and reassure the patient.
- Identify a fire-free escape route.
- Help remove burning material from the patient and place it on the floor to extinguish.
- Help extinguish the fire, if the fire is small.
- Disconnect any electrical equipment that can be disconnected safely.
- Save fire material for later investigation.

Orderlies

- Ensure that hallways, evacuation routes, and exits are unobstructed.
- Help obtain addition supplies, as necessary.

Outside the OR

Charge nurse or clinical leader designated to handle OR fire response

- Determine the number of people in the area.
- Instruct all visitors to exit the area in an orderly fashion.
- Note the time the fire began.
- Communicate with the fire chief or safety officer regarding the fire location.
- Communicate with other ORs as to the status of the fire and plan of action.
- Work with the anesthesiologist with regard to triage priorities for evacuation.
- Determine whether to evacuate.

Administrative assistant

- Follow organization procedure for reporting a fire to the authorities.
- Keep telephone lines open.[17]

Figure 4-1. *Staff members need to be more aware of their roles in a fire emergency and know how to perform them.*

Educating Staff

To reduce the risk of surgical fires, organization staff should be educated on surgical fire prevention strategies. Every member of the surgical team should receive this education, including surgeons, anesthesiologists, nurses, and so forth. Areas to emphasize in training efforts include the following:

- How to minimize fire ignition risks, such as proper equipment use and care
- How to minimize oxidizer risks, including reducing oxygen concentration
- How to minimize fuel risks, such as properly preparing the patient for surgery
- Specific risks for the types of surgery conducted in the organization

☞ **Tip:** It may be helpful to educate nurses and physicians together and encourage them to

Sidebar 4-6. The Dangers of Fire Blankets

Some organizations install wool fire blankets, coated with fire retardants, on the wall of the OR, with the expectation that staff members will use them when they need to smother a fire. Although fire blankets can be helpful in other areas of a health care organization, they present some significant risks in the OR. For example, a fire blanket draped on a patient during an OR fire can burn if oxygen is being delivered to the patient. Fire blankets can also inadvertently push a sharp instrument into a patient or trap a fire next to a patient, causing more severe burns.[8]

discuss the best strategies for surgical fire prevention. By involving the staff in the development of surgical fire protocols, an organization can ensure staff buy-in and participation.[8] ☜

☞ **Tip:** Education for perioperative staff members should review flammability levels of the materials and substances with which staff members work. These substances include not just alcohol-based prepping agents but drapes, towels, gowns, hoods, and masks, as well as the dressings, ointments, equipment, and supplies used during surgery. ☜

The staff should also be educated on how to effectively respond to a fire. Each staff member must know his or her role during a fire and be able to execute that role. Fire response education should include the following:

- How to quickly extinguish a fire
- How to treat the patient
- How to safely remove the patient, if necessary
- How to safely evacuate the OR, if necessary
- How to activate the fire safety systems
- How to prevent the spread of smoke
- How to locate and use fire extinguishers

Sidebar 4-7. Fire Extinguishers

Fire extinguishers are not frequently needed to extinguish OR fires. In fact, fire extinguishers have been used in only three or four of the hundreds of OR fires each year.[10] Although fire extinguishers are not the first solution to an OR fire, they may be necessary to put out fires that engulf a patient or extend beyond a patient into the room.

To ensure proper fire response, organizations should make sure that the fire extinguishers located within the OR are appropriate. CO_2 extinguishers are effective at suppressing fires on the surgical drapes, once the drapes have been pulled off the patient. They release a fog of cold carbon dioxide gas and snow. This smothers and cools a fire without leaving a residue. The cold fog may help minimize any injuries from the fire and is unlikely to injure the patient.[5]

Neither water-based nor dry chemical extinguishers are recommended for use in the OR because they can contaminate the surgical site, and cleanup afterward may be difficult. In addition, water-based extinguishers can result in electric shock to users.[5]

Organizations should make sure that designated fire extinguishers are mounted inside the surgical suite and are inspected monthly and maintained annually. The staff should be trained to know where fire extinguishers are located and how to use them.

- The role of outside firefighting personnel, such as the organization's fire response team or the community fire department

Organizations should conduct surgical fire drills to make sure staff members understand their roles during a fire. Participation in a simulated fire drill can provide more in-depth instruction than learning about fire response in a meeting.

☞ **Tip:** To remind perioperative staff members of what they can do to help prevent surgical fires, safety managers at ambulatory care and

FIRE PLAN AND EVACUATION CHECKSHEET

Name: ____________________

OR#: ____________________

Surgeon: ____________________

Surgical procedure(s): ____________________

Estimated time for completion of the procedures: ____________________

Documentation to accompany patient in case of evacuation

Perioperative nursing record

Intraoperative nursing record

Anesthesia record

Physician order sheet

Names of personnel in the room

Name: ____________	Title: ____________
Name: ____________	Title: ____________
Name: ____________	Title: ____________
Name: ____________	Title: ____________
Name: ____________	Title: ____________

Supplies needed to complete the procedure

____

____

____

Figure 4-2. *This sample check sheet can help account for personnel, supplies, and procedures in the event the OR needs to be evacuated.*

office-based surgical centers can download, print, and mount a fire safety poster in each surgical suite. The free poster is available from ECRI; go to http://mdsr.ecri.org and click on Surgical Fire Poster.

Conducting Fire Drills

A critical aspect of OR fire response is regular fire drills. ECRI recommends that surgical staff take part in special drills focused on the following:

- Properly using OR fire-fighting equipment
- Dealing with patient rescue and escape
- Identifying and locating medical gas, ventilation, and surgical systems and controls
- When, where, and how to shut off these systems
- Using the organization's alarm system and its procedure for contacting the local fire department

After completing drills, it can be helpful to conduct a debriefing session to evaluate the fire response, highlight aspects that went well, and identify areas that require improvement. Drill debriefings can determine the need for more training, the need to revamp response procedures, or the need to tweak certain aspects of fire response, such as the most appropriate door from which to evacuate or the most effective way to extinguish a fire. Some organizations choose to create an evaluation form to "score" a fire drill. This form allows organizations to evaluate a variety of components of the drill, including staff response, rescue efforts, fire containment, and so forth. For more information about fire drills, see Appendix 1.

References

1. Based on an example found in Tooher R., et al.: Surgical fires and alcohol-based skin preparations. *ANZ J Surg* 74:382–385, 2004.

2. Fire marshal bans alcohol-based preps. *OR Manager* 21:5, 7, Jun. 2005.

3. ECRI: A clinician's guide to surgical fires: How they occur, how to prevent them, how to put them out. *Heath Devices* 32:5–24, 2003.

4. Rodgers L.A., Kulwicki A.: Safety in the use of compressed air versus oxygen for the ophthalmic patient. *AANA J* 70:41, Feb. 2002.

5. Pollock G.: Eliminating surgical fires: A team approach. *AANA J* 72:293–297, Aug. 2004.

6. American Society for Healthcare Engineering (ASHE): *Use of Alcohol-Based Surgical Prep Solutions: ASHE Analysis of NFPA 99 and Guidance Articles.* Chicago: ASHE, Mar. 18, 2005.

7. National Fire Protection Association (NFPA):Tentative Interim Amendment to NFPA 99: *Standard for Health Care Facilities,* 2005 edition. NFPA Standards Council, Jul. 29, 2005. http://www.ashe.org/ashe/codes/nfpa/nfpa099_proposeamend_absp.html (accessed Sep. 22, 2005)

8. Wasek S.: Preventing surgical fires. *Outpatient Surgery Magazine* Dec. 2003, pp. 28–35.

9. Salmon L.: Fire in the OR: Prevention and preparedness. *AORN J* 80:42–58, Jul. 2004.

10. Meeting JCAHO's goal on surgical fires. *OR Manager* 20: 26–28, Nov. 2004.

11. Bales L., et al.: Be aware of fuel sources, ignition to reduce fire risk. Supplement to *Same Day Surg* Apr. 2005., pp. 1–3.

12. Highlights of the Fall Executive Committee of the AAP section in *Anesthesia and Pain Medicine.*

13. Holak E.J., et al.: Carbon monoxide production from sevoflurane breakdown: Modeling of exposures under clinical conditions. *Anesth Analg* 96:757–764, Mar. 2003.

14. Woehlck H.J.: Sleeping with uncertainty: Anesthetics and desiccated absorbent. *Anesthesiology* 101:276–278, Aug. 2004.

15. Laster M., et al.: Fire from the interaction of anesthetics and desiccated absorbent. *Anesth Analg* 99:769–774, Sep. 2004.

16. McCarthy P., et al.: Home study program: Fire in the OR—Developing a fire safety plan. *AORN J* 79: 588–600, Mar. 2003.

17. Moyer P.: Operating room fires: How to prevent and minimize spread. *Today's Surg Nurse,* Nov./Dec. 1998, pp. 13–39.

Chapter 5: Ensuring Infection Control and Prevention

It Could Have Happened This Way

William Blake had successful triple bypass surgery and was expected to make a full recovery. He was 51 years old and, other than the need for the surgery, was in good health. As he recuperated in the hospital, an infection began to form around the surgical site. As the days progressed, bacteria that were resistant to antibiotics infiltrated the wound and traveled throughout his bloodstream. Mr. Blake was placed on an escalating series of antibiotics and remained in critical condition in the intensive care unit (ICU). His family was told that he had a 50% chance of surviving the incident. As the days progressed, so did Mr. Blake's infection, ultimately taking his life.*

While appalling in itself, the tragedy of this incident is compounded by the fact that it could have been prevented. Although the infection may have resulted from many factors, including instruments that were not properly sterilized, lack of prophylactic antibiotics, or an unclean environment, the fact remains that the infection most probably could have and should have been avoided.

Health care–associated infections (HAIs) are unnecessary and largely preventable. Surgical site infections (SSI), such as the one discussed above, account for about 40% of all HAIs.[1] If a patient gets an infection in a health care organization, it can have devastating effects on his or her health. Studies show that patients who develop SSIs are twice as likely to die as those who do not. These patients are also 60% more likely to require ICU admission, remain in the hospital twice as long, and have a rate of readmission six times higher than those with no infection.[2] In addition to the effects on patients, health care organizations pay a high price for infections. SSIs can cost health care organizations on average over $3,000 per case. Moreover, the price of treating health care workers who get infections from patients or the environment can also affect an organization's bottom line as workers' compensation, reduced productivity, and staffing costs can all go up.

Infection control in the operating room (OR) involves preventing the patient from getting infections, preventing the patient from giving infections to surgical staff, and preventing the spread of infections to others outside surgery. To achieve optimal infection control in the OR, organizations must consider both environmental factors and clinical factors. This chapter addresses both of these categories and offers tips and suggestions on how to reduce the risks associated with infections in the OR.

Standard IC.1.10. The risk of developing a health care–associated infection is minimized through an organizationwide infection control program.

Addressing Environmental Factors

A safe and clean OR can mean the difference between a successful surgical outcome with a smooth recovery and a long, arduous recovery plagued by infection. It is important to establish and maintain infection control policies that ensure

* All patient and care-provider names in this book are fictitious.

the cleanliness and safety of the OR so that infections are minimized and the spread of infection is prevented. Several environmental factors should be considered in an infection control plan for the OR, including the following:

- Air quality and ventilation
- Effecting OR cleaning
- Effective medical equipment cleaning
- Effective laundry cleaning

AIR QUALITY AND VENTILATION

Heating, ventilation, and air conditioning (HVAC) systems provide circulating air throughout health care facilities, including in and out of the OR. These systems provide temperature and humidity control and filtration of various contaminants. When HVAC systems are not functioning optimally, they can be a source of patient safety and workplace safety hazards. HVAC systems pose three areas of infection control concern:

- ***Pressure relationships.*** Space pressurization is used to ensure that air movement is directed from a clean space into a less clean space. For example, positive-pressure ORs can ensure clean air in the OR at all times.
- ***Air exchange rates.*** Air exchange rates control how many times the air is replaced in the room. The more clean air that is circulated into a space, the less the impact of any contaminants.
- ***Filtration efficiencies.*** Filtration is the primary defense against fungi in ventilation systems. It can take a variety of forms but could include HEPA filters or prefilters that work to remove particles from the air.

If any of these areas are not functioning properly, then an infection risk is present. Conversely, if these areas are all working at optimal functioning, the risks of infection can be severely limited.

To help ensure proper ventilation of the surgical environment, the Centers for Disease Control and Prevention (CDC) developed guidelines regarding infection control and ventilation for ORs. Following are some of the suggestions included in these guidelines:

- Maintain positive-pressure ventilation with respect to corridors and areas adjacent to the OR.
- Maintain at least 15 air changes per hour, of which at least 3 should introduce fresh air.
- Filter all recirculated and fresh air through appropriate filters, providing a minimum of 90% efficiency.
- In rooms not engineered for horizontal laminar air flow, introduce air at the ceiling and exhaust near the floor.
- Keep OR doors closed except for passage of equipment, personnel, and patients and limit entry to essential personnel.[3]

In case an air ventilation system malfunctions, an organization should have a backup system that can be used for emergency ventilation. Such a system could involve portable units for fans or filters. Obviously, backup systems should be used in the short term, while every effort is made to restore the main ventilation system.

☞ **TIP:** To ensure that an HVAC system is designed properly, organizations should involve design professionals who are appropriately credentialed and adhere to specifications contained in state and/or local codes and to guidelines available from the American Society of Heating, Refrigerating and Air-Conditioning Engineers (ASHRAE) and the American Institute of Architects (AIA). ☜

☞ **TIP:** Fans, coils, belts, and filters within an HVAC system all require regular maintenance to function properly. Organizations should establish preventive maintenance, cleaning, and inspection schedules for the HVAC system and stick to them. It is also important to keep accurate, up-to-date records of not only routine maintenance but also requested maintenance and quality management activities, including the use of instrumentation to determine appropriate air change, filtration, and pressure parameters for areas requiring special ventilation.

Using Proper Air Ventilation to Reduce the Risk of Tuberculosis. *Mycobacterium tuberculosis (M. tuberculosis),* the bacterium that causes tuberculosis (TB), is carried through the air in particles called *droplet nuclei.* These bacteria are not transmitted by surface contact. If these droplet nuclei are inhaled, infection can occur in a susceptible individual. The probability that an exposed individual will become infected is proportional to the concentration of the droplet nuclei in the air and the length of exposure.

Because TB bacterium can be transmitted through the air, proper ventilation in the OR can help reduce the risks of transmitting TB to staff and patients. While all the CDC guidelines for properly ventilating the OR apply to TB patients, the CDC also has some additional guidelines, including the following:

- Patients should be intubated in either an airborne infection isolation (AII) room or in the OR. If intubating in the OR, the staff should not allow doors to open until 99% of airborne contaminants are removed.
- Perioperative staff should use, at minimum, a surgical N95 respirator, approved by the National Institute for Occupational Safety and Health (NIOSH) and the Food and Drug Administration (FDA). A hospital may decide to use something other than a N95, such as a powered air-purifying respirator (PAPR). If a person cannot wear a tight-fitting respirator for a medical reason, a PAPR must be provided. In addition, a person with a beard cannot wear a tight-fitting respirator, such as a N95, and must be provided with another respirator, such as a PAPR.
- When anesthetizing a patient with confirmed or suspected TB, staff members should place a bacterial filter between the anesthesia circuit and patient's airway to prevent contamination of anesthesia equipment or discharge of bacteria into the air.
- Staff should extubate a patient and allow for recovery in an AII room.
- If a patient has to be extubated in the OR, staff should allow adequate time for air changes to clean 99% of airborne particles.
- Staff should use portable, industrial-grade HEPA filters temporarily for supplemental air cleaning.
- If possible, staff should schedule TB patients as the last surgical cases of the day to maximize time for removal of airborne contaminants.[3]

Managing Laser Plume. During surgical procedures that use a laser or an electrosurgical unit, the burning of patient tissue creates a smoke by-product. This by-product contains respiratory irritants, mutagens, carcinogens, and pathogens. The smoke or plume can therefore produce upper respiratory irritation and have a potentially negative effect on reproduction. Although there have been no documented transmissions of infectious disease through surgical plume, the potential for generating infectious viral fragments may exist.[4]

Each year, an estimated 500,000 workers, including surgeons, nurses, anesthesiologists, and surgical technologists, are exposed to laser or electrosurgical plume and consequently are exposed to irritants and pathogens within that plume.

Organizations should consider using a mechanical smoke evacuation system with a high-efficiency filter to manage laser plume. This type of system is particularly important in removing the vaporized tissue in the plume of patients infected with human papillomavirus or TB.[3] If such a system is not feasible or effective, then appropriate respiratory protection is required. For assistance in determining the appropriate personal protective equipment (PPE), organizations can consult the NIOSH respirator selection logic found at http://www.cdc.gov/niosh/docs/2005-100/default.html.

A surgical smoke evacuation system suctions surgical plume, filters it, and exhausts it back into the high-efficiency filter. Such systems can be portable or stationary. Stationary systems are integrated into the facility through piping built into the walls. The perioperative team can control the system from the OR and has a variety of exhaust options, including exhausting back into the OR, outside the building, or via the HVAC system. Portable systems house the control, suction, filtration, and exhaust components on a portable unit, such as a cart or table. Some factors to consider when selecting an evacuation system include ease of use, size, noise of equipment, cost, and ease of maintenance.[5] For more information on managing laser plume, see Appendix 1.

CLEANING THE OR EFFECTIVELY

The OR should be one of the cleanest, if not the cleanest, places in the entire health care organization. To ensure that the OR is cleaned effectively every day, the CDC recommends that terminal cleaning of ORs be performed once every 24 hours, even if no procedure has been performed in the room.[3] Terminal cleaning can help decrease microorganisms and contamination risks (particularly contamination of bacteria that are resistant to antibiotics) and may help control the spread of infection to patients.

Terminal cleaning should be performed in surgical procedure rooms and scrub and utility areas, and it involves using a wet vacuum or a single-use mop and an Environmental Protection Agency (EPA)–registered hospital cleaner. The following equipment should be cleaned during terminal cleaning:

- Surgical lights and external traces
- Fixed and ceiling-mounted equipment
- All furniture, including wheels and casters
- Medical equipment
- Handles of cabinets and push plates
- Ventilation faceplates
- Horizontal surfaces, including countertops, autoclaves, and fixed shelving
- The entire floor
- Kick buckets and scrub sinks[7]

To help ensure effective terminal cleaning, some organizations use a checklist that identifies all the surfaces that need to be cleaned in the OR. Staff members, such as perioperative nurses, can use this form to document areas where cleaning is below standard. Information obtained from the instrument can be used to identify and correct problem areas. In addition, environmental services personnel can use the checklist to verify that they have cleaned all the appropriate parts of the OR.[7] An example is provided in Figure 5-1 on page 63.

> **Sidebar 5-1. Human Papillomavirus**
>
> Human papillomavirus (HPV) is one of the most common causes of sexually transmitted infection (STI) in the world. While most forms of HPV are harmless, some have been known to cause certain kinds of cancers.[6] Organizations can prevent the spread of HPV by using a surgical smoke evacuation system.

CLEANING MEDICAL EQUIPMENT EFFECTIVELY

If not cleaned properly, medical equipment (such as surgical instruments, IV poles, and endoscopes) and supplies (such as bed sheets and surgical gowns) can transmit infection from patient to patient, from patient to staff, or from staff to patient. Direct care staff members are not the only employees at risk from infected equipment and supplies; individuals responsible for maintenance and repair of the equipment as well as laundering are also at risk for exposure.

To make sure that OR equipment and supplies are cleaned properly, organizations should have policies and procedures in place that address the following issues:

Environmental Cleaning Monitor

Room	Shelves	Boom	Bed	Equipment	Lights	Overhangs	Ledges	Doors	Monitors	Counters	Furniture	Wall	Ceiling	Floor
1														
2														
3														
4														
5														
6														
7														
8														
9														
10														
11														
12														
13														
14														
15														
16														
17														
18														
19														
20														
21														
22														
23														
24														
25														
26														

RN QA OR Signature:
RN CNOR OR QI Rep:
Anesthesia Technician:
ORA Team Leader Signature:
3rd Shift EVS:
Date:

LEGEND:

Definition of Site:

Shelves—any shelving on which equipment sits
Boom—the ceiling-mounted electrical outlet post
Lights—operating room lights, including mounting and fixtures
Overhangs—pieces of equipment mounted in the ceiling, such as cameras, microscopes, etc.
Ledges—any edge that juts out from the wall or the tops of doors
Doors—the surface of the door, especially around the bottom edges, and in the hinged corner
Monitors—any monitors mounted in the room, including face, mounting, and cables
Counters and tables
Floor—the surface of the floor, including the corners and edges of the room

Definition of Codes:

D—dust accumulation of matter, usually linty in nature, but may appear like a fine powdery substance
D(anes)—dust accumulation on anesthesia machine, tubing, and equipment
B—blood; can be grossly present or can appear as minute spots
S—suture; usually strands of suture material, which can be found on the floor
P—paper; scraps of paper found on the floor during inspection
T—trash, which can be single or multiple scraps

Figure 5-1. *Completing a checklist such as this one can help ensure that all appropriate areas of the OR have been cleaned.*

Source: Pitt County Memorial Hospital, Greenville, NC.

Sidebar 5-2. Dust in the OR

Dust in the OR is never a good thing. It is a vehicle for transporting microorganisms, including human skin and hair, fibers, pollens, mold, fungi, environmental pollution, and components of the room, such as paint, wood, paper fibers, and sheetrock. Circulating air can transport dust that can settle into an open wound, either through direct contact or by settling on equipment surfaces that then transfer dust into the wound. A critical component of ensuring infection control and prevention in the OR is controlling and eliminating dust. Regular and thorough cleaning can address this issue and provide a safe environment for sterile procedures. It is important to note that perioperative nurses should make sure that rooms are free of dust and debris before opening sterile packs and materials.[7]

- Which equipment and supplies must be cleaned as opposed to which are disposable
- When equipment and supplies must be cleaned
- How often equipment and supplies must be cleaned
- How equipment and supplies must be cleaned

Nursing, infection control professionals, housekeeping staff, and biomedical staff should all be involved in developing these policies. Organizations that may not have on-site biomedical or laundry staff that could participate in policy development should nevertheless develop procedures and policies that ensure that, in addition to direct care staff, the people who repair and maintain equipment and clean soiled laundry are protected from contamination.

Medical equipment should be cleaned before and after each patient use as well as when it passes from one department to another. For example, all equipment should undergo appropriate decontamination before reaching an equipment maintenance department and then again before returning to the OR.

There are four types of cleaning that can help remove dirt and pathogens from equipment. Depending on the type of equipment and its intended use, one of the following four methods should be used to clean a piece of equipment:

- ***Cleaning.*** Removes all visible dust, soil, and any other visible material that microorganisms may find favorable for continued life and growth. Scrubbing with hot water and detergent is usually sufficient.
- ***Decontamination.*** Removes disease-producing organisms and thus renders equipment safe to handle
- ***Disinfection.*** Destroys most disease-producing organisms but not all microbial forms. There are three levels of disinfection:
  - *High-level disinfection.* Kills all organisms except high levels of bacterial spores. This is usually done via a germicide cleared by the FDA.
  - *Intermediate-level disinfection.* Kills mycobacteria, most viruses, and bacteria. This is usually done via a chemical germicide registered as a tuberculocide by the EPA.
  - *Low-level disinfection.* Kills some viruses and bacteria. This is usually done with a chemical germicide registered as a hospital disinfectant by the EPA.
- ***Sterilization.*** Destroys all forms of microbial life, including bacteria, viruses, spores, and fungi. There are several types of sterilization, including steam, ethylene oxide, ozone, and flash sterilization. (See pages 65–66.)

More than 30 years ago, E.H. Spaulding[8] devised a classification system for determining the appropriate cleaning strategy for equipment. Organizations might want to use this system as a jumping-off point for classifying their equipment for decontamination. Spaulding classified items for patient care into three categories: critical, semi-critical, and noncritical. These terms refer to the intended use of the device and not the potential degree of contamination. For example, the noncritical category does not imply that items cannot carry contaminants but that their degree of causing harm to health care staff and patients is not critical. These are the three categiories:

- *Critical.* Items in this category must be sterilized. They may include devices used to enter or come into contact with sterile tissues, such as instruments entering a surgical incision or cardiac and vascular catheters, implants, and needles placed into the vascular system.
- *Semicritical.* Items in this category generally require a high level of disinfection. These may include items that come into contact with nonintact skin or mucous membranes, such as respiratory therapy equipment, anesthesia equipment, and flexible endoscopes.
- *Noncritical.* Items in this category require basic cleaning and low-level decontamination. Items that only touch intact skin would fall into this category because the skin acts as an effective barrier to most microorganisms. Items within this category may include crutches, bedboards, blood pressure cuffs, and a variety of other medical accessories, as well as nonmedical accessories, such as recreational equipment.

The Pros and Cons of Different Sterilization Methods. While organizations can sterilize medical equipment in a variety of ways, there are some advantages and disadvantages to different methods. Following is a discussion of some of the different methods and their pros and cons:

- *Steam sterilization.* Steam sterilization is a very common form of sterilization that generally involves heating an instrument in an autoclave and employing saturated steam under pressure. Temperatures in steam sterilization should reach least 121°C (250°F). The critical factors in ensuring the reliability of the steam sterilization method are proper temperature and time and the complete replacement of the air with steam, so there is no entrapment of air.[9] Steam sterilization is not appropriate for all types of equipment, and organizations should consult a manufacturer's guidelines before using steam sterilization on a particular piece of equipment.
- *Flash sterilization.* The Association for Advancement of Medical Instrumentation (AAMI) defines *flash sterilization* as the process designated for the steam sterilization of patient care items for immediate use.[10] Organizations use flash sterilization when perioperative staff members must sterilize instruments quickly, such as when an instrument becomes contaminated or a sterilized instrument is not immediately available. It is important to note that flash sterilization should not be used "for reasons of convenience, as an alternative to purchasing additional instrument sets, or to save time."[11]

Items that are flash sterilized are placed unwrapped in a gravity displacement or prevacuum sterilizer and steam sterilized at high temperatures. These items are wet and hot when they are removed from the sterilizer. Hot, wet, sterile items are subject to contamination when transported uncovered from the sterilizer to the sterile field. In addition, as a result of rushing to get a piece of equipment sterilized, some items may be put in the sterilizer without being adequately cleaned. This may result in insufficient sterilization. If sterilization is insufficient, then a potentially nonsterile instrument can be used during surgery, increasing the risk for infection and resulting in an adverse outcome.[12] Flash sterilization use should be limited within an organization. The Association of periOperative Registered Nurses (AORN) recommends that it should be used only in carefully selected clinical situations when certain parameters are met. These parameters include when proper work practices for cleaning, decontamination, inspecting, and arranging instruments in the sterilizing tray or containers are followed; when the physical configuration of the department or work area provides for direct delivery of sterilized items; and when standard procedures for personnel safety and aseptic handling of sterilized instruments during transfer to the point of use are followed and audited.[13]

■ ***Ethylene oxide sterilization.*** This is a sterilization process that involves the chemical ethylene oxide (EtO). EtO is a common chemical, and nearly half of all medical devices sterilized are sterilized using EtO.[14] EtO is very effective in killing microbes in hard-to-reach and hard-to-clean places. It can thoroughly clean an instrument without damaging it. It does require significant aeration time, however, to preserve the safety of the individuals handling the equipment. Exposure to EtO is very harmful to people, and the Occupational Safety and Health Administration (OSHA) has specific standards that govern its use (see OSHA's Web site, at http://www.osha.gov). At toxic levels, EtO smells like ether. Exposure to EtO in health care settings is usually through inhalation, and it can cause a variety of symptoms, ranging from eye pain and sore throat to convulsions, nerve damage, and impaired thinking. In addition, staff members exposed to EtO have an increased chance of contracting some cancers. Certain forms of EtO can be extremely flammable, so organizations should consider using blended forms that have less of a fire risk.[14]

■ ***Ozone sterilization.*** This is a chemical sterilization process that uses ozone, which is a combination of three atoms of oxygen rather than the normal two. It works by establishing negative pressure in the sterilization chamber. Ozone is produced in the sterilizer itself from the water oxygen and electricity. At the end of sterilization, the ozone is converted back to oxygen and water vapor, which is released into the room. Operationally, ozone sterilization requires electricity, some potable water that has passed through a demineralizing filter, and medical-grade oxygen gas. As with all other chemical sterilants, excessive exposure to ozone can cause a serious health hazard. Fortunately, ozone is highly detectable by the human nose, and a staff member would know if ozone was present in the area before it became a health hazard.[15]

■ ***Gas plasma sterilization.*** This method uses a combination of hydrogen peroxide vapor and low-temperature gas plasma to rapidly sterilize medical instruments without leaving toxic residues. This method is particularly suited to heat-sensitive and moisture-sensitive instruments.

Regardless of the sterilization method organizations choose, they should have policies and procedures in place that govern the safe use of these methods. Staff should be familiar with these policies and consistently comply with them.

☞ **Tip:** Implantable medical devices should never be flash sterilized. AAMI recommends that if an implantable device is sterilized on site, the perioperative staff should quarantine the device and await the outcome of biological monitoring.[14] ☜

Ensuring That All Equipment Is Cleaned. When designing policies and procedures for OR equipment cleaning, organizations must make sure that such policies and procedures apply to all equipment within the OR, including equipment not owned by the organization, such as demonstration, substitute, and loaner or rental units. Because such equipment moves from organization to organization and is exposed to an unknown variety of potentially infectious agents, safe practices must include appropriate cleaning of equipment before it enters and after it exits the organization. Once policies have been created, organization staff, including nurses, infection control professionals, and biomedical personnel should be educated on the policies.

☞ **Tip:** Organizations can help remind the staff of what equipment should be cleaned, by what methods, and at what time, by means of noticeable, easy-to-read labels on the equipment. Checklists can be used to ensure that the appropriate staff members follow all the procedures necessary to effectively clean the equipment. Logbooks, which record the performance of decontamination procedures, should be available and regularly monitored to ensure that proper cleaning procedures are performed. ☜

☞ **TIP:** Certain types of equipment are more difficult to clean effectively than others. For example, laryngoscopes blades and handles can be particularly challenging to clean. Although there are a variety of information sources regarding how to effectively clean such equipment, the American Society of Anesthesiologists (ASA) provides some helpful tips that can be applied to other instruments posing difficulties:

- Clean the instruments with a low-sudsing enzymatic detergent as soon as possible after use.
- Use disposable brushes for cleaning or make sure that brushes are cleaned, sterilized, and disinfected daily.
- When using liquid disinfectant, all internal and external surfaces should be in contact with the disinfectant for at least 20 minutes, at which time the equipment should be rinsed with sterile water followed by a 70% alcohol rinse.
- Nonimmersible parts should be physically cleaned with water and detergent and then wiped with 70% alcohol.[16]

As always, it is important to follow the manufacturer's cleaning instructions. ☜

Cleaning Endoscopes and Bronchoscopes. Water, although essential in cleaning medical equipment, can also be a source of microorganisms that cause infection. Recently, there have been several reports regarding infected endoscopes and bronchoscopes that have resulted from inadequate cleaning techniques.

To prevent the inadvertent contamination of endoscopes and bronchoscopes, organizations should clean, disinfect, and maintain automatic endoscope reprocessing equipment according to the manufacturer's instruction and relevant scientific literature. According to the CDC, organizations should follow specific guidelines when cleaning and disinfecting endoscopes and bronchoscopes:

- To rinse disinfected endoscopes and bronchoscopes, the staff should use water of the highest quality appropriate for its practice. This could include sterile water or bacteriologically filtered water (water filtered through 0.1 to 0.2 micron filters).
- After rinsing endoscopes and bronchoscopes, the staff should dry internal channels, using a proven method, such as by flushing with 70% to 90% alcohol solution or by using forced air. This drying will lessen the potential for proliferation of waterborne microorganisms and help prevent biofilm formation.[3]

Organizations should have policies that explicitly cover how the staff must clean, disinfect, sterilize, and store endoscopes and bronchoscopes. Within these policies organizations should consider requiring endoscopes and bronchoscopes to be cleaned and sterilized after each procedure and before the first patient of the day. Organizations should monitor for strict adherence with any reprocessing protocols.

As with any other kind of cleaning and sterilization process, there are always new technologies emerging. Organizations should keep abreast of these technologies and consider them when revising policies and procedures.

Preventing Cruetzfeldt-Jakob Disease (CJD). Surgical devices used during surgery are oftentimes reused on other patients after being sterilized. While this is not usually a problem, it is if a patient has CJD and the instruments used in his or her surgery are reused on someone else. CJD is a rare, fatal, dementing brain disorder in which a person's brain disintegrates. There is currently no known cure. Because of its rarity, traditional sterilization methods may not always get rid of the CJD infectious agents known as *prions.* Prions are resistant to a number of standard disinfection and sterilization methods.

An additional concern about CJD is that it is often not immediately detectable and may take years to identify. Consequently, if there is any chance that a patient could have this rare brain disorder, organizations should consider disposing of any surgical instruments used on the patient after use. Organizations should also consider using disposable medical devices for brain surgery.

If organizations choose to use reusable devices, they should be cleaned using sodium hydroxide and a high-temperature autoclave or quarantined until a biopsy confirms that the patient is not infected with CJD.[17] Another advisable cleaning method involves soaking devices in concentrated sodium hydroxide for one hour. It is important to note that sodium hydroxide is extremely corrosive.

☞ **TIP:** In cleaning the room of a surgical patient who has CJD, an organization should consider whether any contamination by central nervous system tissue has occurred. If not, extraordinary cleaning measures are not necessary. However, if central nervous system or cerebral spinal fluid has contaminated the OR, housekeeping staff should use a sodium hypochloride (bleach) solution with approximately 10,000 to 20,000 ppm available chlorine. It is important to note that the routine diluting (1 to 10) of hypochloride yields only 5,000 ppm. To obtain the necessary dilution, organizations must dilute 1 to 4 or 1 to 5. This chemical should have contact with the contaminated surface for 30 minutes to an hour and then be blotted up with an absorbent material. The area should be rinsed and treated with water. The absorbent material should be disposed of in appropriate waste containers. ☜

CLEANING LAUNDRY EFFECTIVELY

The OR generates a great deal of laundry. As previously mentioned, linens, scrubs, and patient gowns can all be contaminated and require cleaning. All organizations should have a way to separate dirty linen and scrubs from clean linen. To remove pathogens from soiled laundry, such as bed sheets and gowns, the Association for Professionals in Infection Control and Epidemiology (APIC) recommends that laundering should be done for a minimum of 25 minutes in at least 160 degree water or with chlorine bleach. Laundry staff members should wash their hands after handling contaminated laundry and should not be permitted to eat, drink, or smoke in the workplace.

When storing clean linen and supplies, it is recommended that they be kept at least six inches off the floor to prevent contamination by floor mopping. Even when linens are enclosed in plastic wrap, splash and splatter from mopping activities can contaminate the exterior surface and pose an infection risk to staff members who handle them.

Sidebar 5-3. The Barrier Standard for Surgical Gowns and Drapes

When choosing surgical gowns for the staff, organizations should give consideration to how those gowns will protect staff members and patients from the spread of infection. The safety and performance of a surgical gown or drape are dependent on the materials from which it is made and also its design. There is considerable variation in design among the variety of protective apparel and drapes currently available.[18]

AAMI recently released guidelines to help organizations select surgical gowns, protective apparel, and drapes. Some factors the guidelines address include the following:

- The types of materials used in the construction of protective apparel and drapes
- The safety of the protective apparel and drapes
- The performance characteristics of protective materials
- What to consider in choosing the level of barrier performance needed for expected exposure risks
- The appropriate care of protective apparel and drapes
- Proper disposal of used protective apparel and drapes

AAMI recommends that organizations screen products based on material and product tests. The guidelines offer general relationships between barrier performance and expected exposure risks.[18] For full access to the AAMI guidelines, go to http://www.aami.org.

Addressing Clinical Factors

By keeping the environment clean, an organization can do much to reduce the chance of infection. However, organizations can take several clinical steps to address the issue as well. Choosing and using effective antiseptic agents, implementing initiatives to reduce SSIs, ensuring staff health, and adequately preparing for an IC emergency can all go a long way toward controlling infection and preventing its spread.

Determining the Effectiveness of Antiseptic Agents

One of the main contributing factors to infections in the OR involves contamination by micro-organisms located on a patient's skin. Preooperative skin antisepsis, through the use of an antiseptic agent, helps eliminate these micro-organisms and is critical to the prevention of infections in the OR. Some of the current antiseptic agents available in the United States include the following:

- Isopropyl alcohol
- Povidone-iodine
- Chlorhexidine combine with isopropyl alcohol
- Chlorhexidine alone

When choosing which antiseptic to use, organizations should consider a number of factors, including the efficacy of the antiseptic, the type of procedures for which it will be used, the safety issues surrounding the product, and the effects it will have on the staff using it. Organizations should do careful research and determine a consistent antiseptic to use. As mentioned in Chapter 4, some antiseptic agents can pool on the patient's skin and represent a surgical fire hazard. With any antiseptic, organizations should train staff on how to properly use it to prevent adverse events, such as surgical fires.

☞ **Tip:** An antiseptic that is made up of a combination of two antiseptics with different mechanisms of action can be better at fighting bacteria than a single antiseptic. Organizations should consider this when choosing an antiseptic.[19] ☜

Reducing Surgical Site Infections

SSIs continue to be a major problem for many surgery patients. As previously mentioned, SSIs account for nearly 40% of all HAIs. The following four variables determine whether a surgical wound will heal normally or become infected[20]:

1. The amount of bacteria that enters the wound area. This can be from the air or via contaminated medical instruments.
2. The virulence of the bacteria. The more virulent the bacteria, the more likely an infection will occur.
3. The microenvironment of the wound. Dead tissue can harbor bacteria, and foreign bodies can also house microbes and promote infections.
4. The strength of the patient's defenses against infection. Some patients have better responses to bacteria than others. Outside influences, such as shock, chronic illness, malnutrition, hypothermia, and hyperglycemia, can all reduce resistance to bacteria.

The good news about SSIs is that with proper precautions, they can be prevented. The following sections discuss some of the proactive strategies organizations can take to reduce and eliminate SSIs within their structures.

Using Prophylactic Antibiotics. Appropriate use of prophylactic antibiotics plays a critical role in preventing SSIs. The effective use of prophylactic antibiotics requires three key factors:

1. ***The appropriate selection of the antibiotic.*** When selecting the appropriate antibiotic, the selection will vary depending on the type of surgery. The American Society of Health-System Pharmacists (ASHP) and the CDC have developed guidelines for appropriate antibiotic selection. When developing antibiotic prescribing protocols, organizations should review and consider these guidelines. In

addition, organizations should consider physician input and their formularies. By standardizing the process, organizations can remove the need to write a unique antibiotic order for every patient. This can prevent prescription errors or omission. A good protocol will contain clear guidance for the staff as to appropriate substitutions in case of allergies or when a physician should be contacted for a different order.[21]

2. ***The correct timing of the first dose.*** Studies have shown that the timing of the first dose of prophylactic antibiotics plays a crucial role in the risk of SSI. Ideally, the first dose should be given before the incision but no more than 60 minutes prior. If the dose is given too early, its peak effect may have already passed before the incision is made. If the antibiotic is given at the time of incision or after, the peak effect may occur too late to provide optimum protection. When giving the first dose, organizations should use rapid infusion so the start time of the dose is very close to the end time. If the patient requires a blood pressure cuff during surgery, and the cuff is to be applied near the surgical site, the first dose should be fully infused prior to cuff inflation; otherwise, the cuff may prevent the full dose from reaching the site.[21] When determining the dose of an antibiotic, the staff should make sure to adjust the dose for the patient's weight. A large patient may require a higher dose. Also, for longer surgeries, multiple doses may be necessary. To ensure the appropriate timing of antibiotic administration, organizations should designate responsibility for administration. This responsibility could go to anesthesiologists, circulating nurses, or preoperative nurses. Someone should also be designated to document the dose and incision time. Thus, organizations can review their timing and determine whether improvements must be made.

3. ***The appropriate discontinuation postoperatively.*** Depending on the type of surgery, organizations may, by policy, discontinue antibiotics immediately after surgery ends. For other types of surgery, such as orthopedic surgery, antibiotic use may continue postoperatively, but not for more than 24 hours. In some limited situations, extended antibiotics may be given. For example, cardiac surgery warrants the use of antibiotics for 48 hours after surgery ends. Again, protocols and standard order sets can help ensure that antibiotics are stopped at the appropriate time. Organizations may want to standardize timing within these protocols and require physicians to opt out of their use. In other words, in these situations, a patient will automatically receive antibiotics for a predetermined time unless the physician indicates that he or she would like the patient to receive the antibiotic for a longer period.[21]

Once organizations create their prophylactic antibiotic policies and protocols, they must educate practitioners on the appropriate use and administration of antibiotics. Some suggestions on how to do this include the following:

- Provide educational resources, such as articles and training sessions, on the various antibiotics that are appropriate for different types of procedures. Provide mandatory in-services about proper use of antibiotics in surgical situations.
- Provide a prophylactic antibiotic forcing function, such as a dedicated nursing professional who screens all surgical patients for prophylactic antibiotics prior to surgery. This individual would determine a patient's candidacy for antibiotics and make sure that the proper medications are given.
- Remind the staff about antibiotic protocols through tent cards, posters, note pads, checklists, and so forth.

☞ **TIP:** To help ensure proper timing of antibiotics, organizations should consider standardizing the process as much as possible. This might include storing antibiotics in ready-to-administer doses in the OR.[21] ☜

Removing Hair Appropriately. In years past, nurses would prep patients for surgery by shaving them. Even today, some organizations still use razors to shave hair around the intended surgical site. However, several studies have shown that patients whose hair was not removed before surgery had a significantly lower rate of infection than those who did have their hair removed.

When creating a policy for appropriate hair removal, organizations should consider recommending that hair be removed only when necessary. If necessary, staff members should use clippers rather than razors to remove hair. Razors should ideally not even be present in the organization, and the purchasing and supply departments should be asked not to reorder any razors.

While this may seem like an easy switch, many organizations struggle with it. Staff members are used to shaving patients and may have a hard time switching their practice. Organizations should educate the staff about the importance of clipping versus shaving and provide training on how to properly use clippers. Patients, too, should be educated on the importance of not shaving before surgery, especially near the planned surgical site.

Ensuring Glycemic Control. Higher levels of glucose can increase the risk of SSIs. Organizations can maintain a patient's glucose levels by administering insulin after surgery. There are several protocols in existence that can help an organization determine a sliding scale of insulin administration. For example, many hospitals use the Portland Protocol developed at Providence Heart and Vascular Institute in Everett, Washington, to help dose insulin based on serum glucose levels.[22] Instead of using a sliding scale, organizations may want to create a protocol that requires the infusion of insulin for 72 hours postoperatively, to maintain blood sugars.

Preventing Hypothermia. One of the most common complaints from people who are having surgery is being cold. Not only is this situation uncomfortable and unnecessary, it has also been shown to contribute to an increased risk of complications, including SSI.[23] To help prevent hypothermia, organizations may use a combination of blankets, forced-air patient warming systems, warmed intravenous (IV) fluids, or slightly elevated room temperatures. It is also important to have consistent equipment and methods for measuring and monitoring patient temperature.[21]

Providing Supplemental Oxygen During Surgery. In addition to the proper use of prophylactic antibiotics, maintaining high levels of inspired oxygen during surgery can help reduce the likelihood of SSI. To routinely accomplish this activity, organizations should assign responsibility for documenting oxygen levels, design and implement protocols and standing orders for supplemental oxygen, and design a training program for the staff that outlines the importance of this initiative.

☞ **Tip:** Reducing SSIs should be a team effort in a health care organization. Organizations can benefit from forming a multidisciplinary team to explore the issue, determine goals, develop protocols, and spearhead improvement. Possible members of a multidisciplinary SSI improvement team could include representatives from pharmacy, OR nursing, preoperative nursing, postoperative nursing, anesthesia, infection control, surgeons, and other members of the medical staff. ☜

Ensuring Staff Health

Every year the flu season causes millions of people to get sick. Health care staff are not exempt from this and can develop infections that, untreated, can cause significant risks to their patients. One way to combat these risks is to have strict policies in place that outline when an employee can and, more importantly, cannot report for work. For example, if an employee has an elevated temperature or infected wound, policies should dictate that he or she stays away from work until the temperature is normal for 24 hours or the wound heals.

In addition to employee health policies, organizations should consider vaccinating the staff for preventable communicable diseases, such as influenza. This can limit the number of health care

Sidebar 5-4. A Classification of Wounds

While several factors can contribute to the likelihood of an SSI, the type of wound can also be a factor. Following are descriptions of four types of surgical wounds:

- ***Clean wound.*** These are uninfected operative wounds, where the operative procedures do not enter viscus, respiratory, alimentary, or genital or urinary tracts. Examples of the types of procedures that cause this type of wound include breast biopsies and open-heart surgeries. SSI rates for this type of wound tend to be less than 2%.
- ***Clean-contaminated wound.*** These are operative wounds according to which respiratory, alimentary, or genital/urinary tracts are entered but under elective and controlled circumstances. Examples of procedures that cause this type of wound include gynecological procedures, head/neck cancers that involve the mouth, and pulmonary resection. SSI rates for this type of wound tend to be 4% to 10%.
- ***Contaminated wound.*** These are wounds where gross contamination is present but with no obvious infection. This includes operations where major breaks in aseptic technique have occurred. Examples of these include gunshot wounds to the abdomen and accidental bowel perforations. SSI rates for these wounds are greater than 10%, even with preventive antibiotics and other strategies.
- ***Dirty wound.*** These are wounds that involve an active infection that is already present. Procedures that lead to these kinds of wounds include abdominal exploration of peritonitis and the exploration and draining of abscesses. As with contaminated wounds, the SSI rates for these types of wounds are greater than 10%, even with preventive antibiotics and other strategies.[24]

professionals who get sick during a particular season and can thus reduce the potential for the spread of infections.

Responding to an Infection Control Emergency

The nightly news regularly shares disturbing facts on the lack of preparedness in the United States for infection control emergencies, such as the avian flu, a bioterrorism attack, or West Nile Virus. To be prepared for such disasters, organizations should review their emergency management plans and determine how to address influxes of infectious patients. While the scope of this publication does not allow an in-depth discussion of how organizations should prepare for an infection control emergency, following are some suggestions organizations should consider when preparing ORs for such a disaster:

- ***Limit traffic in the OR.*** If an organization experiences an infection control emergency, such as an outbreak of avian flu, the organization should consider ceasing all elective surgery. While some urgent procedures must continue, the organization should delay them, if possible, until a true understanding of the nature of the outbreak is reached.
- ***Designate certain ORs for use by infected patients.*** By keeping the infected patients segregated to certain ORs and designating specific routes and elevators for patient transport, organizations can limit the exposure of healthy patients to infection control risks. Ideally, these ORs should have separate ventilation systems that can be operated under positive pressure and have higher-than-normal air exchanges.
- ***Monitor the health of staff.*** If staff members show any signs of infection, they should be quarantined. In addition, any patient contact with an infected staff member should be documented so that any isolation measures can be implemented quickly. Before taking any actions, organizations should consult their employee health services or infection control professional for specific instructions.

■ ***Use disposable medical equipment, when possible.*** If it is not possible to use disposable equipment, then reusable equipment should be disinfected especially thoroughly. During an infectious disease outbreak, an organization should consult its infection control professional for specific guidance on disinfection. Disposable equipment should be double-bagged for safe disposal. Staff members handling medical equipment should wear the appropriate PPE. An organization should consult its infection control professional for proper selection and instruction on PPE. In addition, an organization might want to consider sterilizing equipment multiple times to ensure contaminant elimination.

References

1. MedQIC: *Surgical Care Improvement Project: Making Surgery Safer Project Overview.* http://www.medqic.org/scip/scip_homepage.html (accessed May 10, 2006).

2. Kirkland K., et al.: The impact of surgical-site infections in the 1990s: Attributable to mortality, excess length of hospitalization, and extra costs. *Infect Control Hosp Epidemiol* 20:725–730, Nov. 1999.

3. Centers for Disease Control and Prevention: *Guidelines for Environmental Infection Control in Health-Care Facilities,* 2003. CDC June 2003, http://www.cdc.gov/ncidod/dhqp/gl_environinfection.html (accessed May 1, 2006).

4. Occupational Safety and Health Administration: *Laser/Electrosurgery Plume.* http://www.osha.gov/SLTC/laserelectrosurgeryplume/index.html (accessed May 1, 2006).

5. Ball K.: No smoking in the OR. *Outpatient Surgery* 4:62–67, Nov. 2003.

6. National Institutes of Health: *Human Papillomavirus and Genital Warts.* http://www.niaid.nih.gov/factsheets/stdhpv.htm (accessed May 1, 2006).

7. Neil J.: Environmental surveillance in the operating room. *AORN J* 82:43–49, Jul. 2005.

8. Spaulding E.H.: Chemical disinfection of medical and surgical materials. In Lawrence C.A., Block S.S. (eds.): *Disinfection, Sterilization and Preservation.* Philadelphia: Lea and Febiger, 1968.

9. Berkeley Lab Environmental Health & Safety Division: *Sterilization.* http://www.lbl.gov/ehs/biosafety/Biosafety_Manual/html/sterilization.shtml (accessed May 10, 2006.)

10. Gardner M.: Flash sterilization: A questionable practice requires proper usage. *Infection Control Today* http://www.infectioncontroltoday.com/articles/241instrument.html (accessed May 5, 2006).

11. Centers for Disease Control and Prevention. *Guideline for Prevention of Surgical Site Infection, 1999.* http://www.cdc.gov/ncidod/dhqp/pdf/guidelines/SSI.pdf.

12. Leonard Y., et al.: Evaluating use of flash sterilization in the OR with regard to postoperative infections. *AORN J* 83: 672–680, Mar. 2006.

13. Association of periOperative Registered Nurses (AORN): Recommended practices for sterilization in perioperative practice settings. *Standards, Recommended Practices, and Guidelines.* Denver: AORN, Inc. 2005, pp. 459–460.

14. Conviser S.: The future of ethylene oxide sterilization. *Infection Control Today* http://www.infectioncontroltoday.com/articles/061feat4.html (accessed May 1, 2006).

15. Reichert M., et al.: Is the new ozone sterilizer right for your OR? *OR Manager* 19:23–24, Nov. 2003.

16. McKee K.: Tips for preventing equipment cross-contamination: Infection control experts share their advice for keeping three hard-to-clean instruments pathogen-free. *Outpatient Surgery* 3(8):41–46, Aug. 2002.

17. Klein A.: Reusing surgical tools may be fatal. *Contra Costa Times* Dec. 31, 2005.

18. New document feature: Protective barrier guidance. *AAMI* 41:1, 4–5, Jan. 2006.

19. Hibbard J.: Analyses comparing the antimicrobial activity and safety of current antiseptic agents: A review. *J Infus Nurs* 28:194–207, Jun. 2005.

20. Odem-Forren J.: Surgical site infection: Still a reality. *OR Insider* Nov. 2005, pp. 16–20.

21. Griffin F.: Preventing surgical site infection. *Nurs Manag* 36:20–26, Nov. 2005.

22. Providence Health System: *The Portland Protocol.* http://www.providence.org/Oregon/Health_Resource_Centers/Heart_Disesase_Center/PortlandProtocol.htm (accessed May 10, 2006).

23. Kurz A., et al.: Periopertaive normothermia to reduce the incidence of surgical-wound infection and shorten hospitalization. *New Engl J Med* 334:1209–1216, 1996.

24. Odem-Forren J.: Watch that incision! Thwart surgical site infections with antimicrobial prophylaxis. *Nurs Manag* 35:16–21, 2004.

Chapter 6: Maintaining Proper Hand Hygiene in the OR

It Could Have Happened This Way

During her annual exam, Anne Somerville's gynecologist discovered a small lump in her left breast. The physician ordered a mammogram, which showed a small cyst. Although reasonably confident the cyst was benign, the doctor suggested that Ms. Somerville have the cyst removed and biopsied. It was a routine procedure that the hospital performed multiple times a day. There was very little risk involved, and Ms. Somerville went to the hospital with the full expectation that she would be back to her regular routine in a couple days. Ms. Somerville had three young children and could not afford to be out of commission for too long.*

The surgery itself was uneventful. However, shortly after returning home from the hospital, Ms. Somerville's left breast began to hurt. The area around the surgical wound was red, slightly swollen, and hot to the touch. She began running a low-grade fever. Ms. Somerville returned to her physician, who diagnosed an infection, and prescribed a course of oral antibiotics. Within 24 hours, Ms. Somerville's fever had elevated, and the pain associated with the infection was worse. Ms. Somerville returned to her physician who admitted her to the hospital for intravenous (IV) antibiotics. Ms. Somerville spent three days in the hospital recovering from her infection before being sent home. It took her another full week to return to her normal activities. The hospital spent $1,500 treating Ms. Somerville's infection.

The hospital reviewed Ms. Somerville's case to see what could have caused her infection. After ruling out several possibilities, the organization determined that the alcohol-based surgical scrub normally used for hand scrubbing had been temporarily unavailable on the day of Ms. Somerville's surgery, and staff members had been scrubbing using plain soap and water. In addition, several of the surgical staff members admitted to not having washed their hands after using the bathroom.

* All patient and care-provider names in this book are fictitious.

As in the example in the Chapter 5, Ms. Somerville's infection could have and should have been prevented. In this case, proper hand hygiene could have saved Ms. Somerville from significant pain, worry, and inconvenience, and it could have saved the hospital a significant amount of money.

The issue of hand hygiene is by no means new. Nearly 140 years ago, Dr. Ignaz Semmelweis discovered a link between increased hand washing and reduced infection. Despite that discovery, the record of poor hand washing in health care organizations has continued throughout the years.[1]

Recommended Hand-Hygiene Practices for the OR

When it comes to hand hygiene, and specifically hand hygiene in the operating room (OR), several national health care agencies, including the Association of periOperative Nurses (AORN) and the Centers for Disease Control and Prevention (CDC), have developed best practices. While the scope of this publication does not allow an in-depth discussion of all those practices, following are some of the most common ones. Organizations should consider these practices when developing policies and procedures for surgical hand antisepsis:

- All surgical personnel should have short fingernails and refrain from wearing artificial nails in the OR. According to the CDC, fingernails should be kept within one-quarter inch long.
- Before beginning hand antisepsis, staff members should remove debris from underneath their fingernails, using a nail cleaner under running water.
- Surgical staff should not wear jewelry in the OR. All rings, watches, and bracelets should be removed before starting hand antisepsis.
- The staff should use a Food and Drug Administration (FDA)–compliant surgical hand antiseptic for cleaning hands before and during surgery. This antiseptic can include either an antimicrobial soap or an alcohol-based hand rub (ABHR) with persistent activity.
- An organization should have a standardized hand scrub procedure for all ORs, and the staff should be aware of and consistently follow that procedure. There are three main options for this procedure:

1. ***Using an FDA–approved antimicrobial soap in a traditional scrubbing process.*** This process involves washing with regular soap first and then using the antimicrobial soap. A soft, nonabrasive sponge may be used for cleaning with the antimicrobial soap. AORN does not recommend using a brush for cleaning, as brushes can lead to skin damage and an increased shedding of skin cells and bacteria.[2] When using an antimicrobial soap, staff should scrub hands and forearms for the manufacturer's recommended length of time. This is usually between two and six minutes.
2. ***Using an FDA-approved ABHR in conjunction with soap and water.*** This process usually involves prewashing hands and forearms with a nonantimicrobial soap and drying hands and forearms completely. After applying the alcohol product, according to the manufacturer's instruction, the staff should allow hands and forearms to dry thoroughly. Repeated applications may be warranted.
3. ***Using an antimicrobial agent for brushless scrubbing.*** Brushless scrubbing involves cleaning the hands and forearms with an antimicrobial agent without the use of a brush or sponge; the person's hands, instead of the brush or sponge, act as the friction agent. It is a relatively new concept that is significantly different from the traditional surgical scrubbing approach. (See Sidebar 6-1, above.)

Sidebar 6-1. Brushless Scrubbing

Recent evidence shows that brushless scrubbing may be as effective as, or even more effective than, the traditional scrubbing process. In addition, brushless scrubbing is believed to be less abrasive than traditional scrubbing. This is important because compromised skin integrity can sometimes increase the chances of bacterial colonization that can lead to infection.

The following steps outline the brushless scrubbing process:

1. Use a nail pick to clean under fingernails. After discarding the nail pick, wet hands and arms with water.
2. Dispense brushless product into both hands, using a foot pump.
3. Rub both hands together, then move up the arms, extending past the elbows. Adding more water will create more lather. Continue to rub hands and forearms for 90 seconds.
4. Rinse hands and arms thoroughly.
5. Repeat steps 2 through 5 for a second application.
6. Dry hands using an aseptic technique.[3]

Antiseptic Product Options

Antiseptics used for surgical hand antisepsis should be fast acting and long lasting. They should remove dead skin, reduce microorganisms on intact skin, provide persistent protection against bacteria, and be effective against a wide variety of

Sidebar 6-2. The Joint Commission Perspective on Hand Hygiene

The Joint Commission has turned its focus to the issue of hand hygiene through its standards and National Patient Safety Goals. These require organizations to improve hand hygiene among staff and comply with the CDC's hand-hygiene guidelines. A complete listing of the CDC's guidelines is available at http://www.cdc.gov/handhygiene/. This report is extremely detailed and well documented. The specific recommendations referred to in National Patient Safety Goal 7 are on pages 31 through 34 of the report.

Each of the CDC hand-hygiene recommendations is categorized on the basis of the strength of evidence supporting the recommendation. To comply with National Patient Safety Goal 7A., all Category I recommendations (including Categories IA, IB, and IC) must be implemented. Category II recommendations should be considered for implementation but are not required for accreditation purposes.

National Patient Safety Goal 7A.
Comply with current CDC hand-hygiene guidelines.

microorganisms. In addition, they should be nonirritating to prevent skin reactions.

There are many different options available for antiseptic hand hygiene, including povidone–iodine solutions, chlorhexidine, and antimicrobial soap. The different options have different benefits. For example, using antimicrobial soap is a more traditional approach to hand hygiene, and the staff may be familiar and comfortable with its use. The downside, as previously mentioned, is that such soap can dry out skin and cause irritation that can lead to bacteria growth and infection. In addition, antisepsis using antimicrobial soap can take more time, use more water, use more laundry, and cost more money.

Alcohol has bacteria-fighting properties that help eliminate bacteria faster than soap can. In addition, bacteria take longer to reappear on skin after the use of an ABHR. A 60% to 95% alcohol preparation provides the most rapid and effective reduction of microorganisms on the skin. However, this type of product does not have a long-lasting effect. Using an ABHR that has an antimicrobial agent in it, such as chlorhexidine, may be preferable for some organizations,[2] as this combination has been shown to be effective in reducing bacteria and does not cause negative skin reactions.[4]

Yet another study showed that preoperative hand decontamination with povidone–iodine enabled sufficient cleaning, provided it was conducted for at least three minutes. This brushless method was effective for shorter surgeries.[5]

Which products an organization selects depends a great deal on the organization, staff preferences, budget considerations, and so forth. It is recommended that an organization form a multidisciplinary committee to give input on the selection process. Different disciplines have different perspectives on the issue, and everyone's opinion should be considered.

Every day, new research emerges on new types of products. Organizations are wise to keep abreast of this information and regularly review their hand-antisepsis processes.

Improving Hand-Hygiene Compliance

While hand-hygiene compliance rates throughout a health care organization can be surprisingly low, hand antisepsis in the OR has been held up as the epitome of how consistent and standardized hand-hygiene processes can reduce infections. Nevertheless, some organizations have found that compliance issues still exist in the OR. For example, surgical staff members may not scrub for as long as required by the organization's protocols, or they may avoid aspects of the hand-antisepsis process. One study showed that a significant portion of the surgical staff that used the toilet in the changing room of the operative theater failed to wash hands afterward.[6]

There are several ways organizations can encourage compliance with OR hand-hygiene protocols. Following is a discussion of some of those methods.

MAKING GOOD HAND HYGIENE PART OF THE CULTURE

Organizations that emphasize good hand hygiene as part of an overall emphasis on safety can help improve hand-hygiene compliance in the OR. By committing to improved hand hygiene at all levels of an organization, an organization can make hand hygiene part of its overall safety focus. Leadership support is critical in this effort because staff pay attention to what organization leadership says and does. Identifying role models in the physician and nursing communities may help improve compliance also. By watching a peer follow appropriate hand-antisepsis protocols, staff members can be encouraged to improve their hand-hygiene habits as well.

PROVIDING EDUCATION TO STAFF

Education programs should discuss the organization's policies, procedures, and protocols regarding proper hand antisepsis. In addition, efforts should discuss the proper use of gloves with regard to hand antisepsis and the need to reclean hands during long surgeries.

Before designing education efforts, it may be helpful to survey staff members about their awareness of the importance and proper methods of hand hygiene. This can help focus education efforts and address areas where staff knowledge is deficient.

CONVENIENTLY LOCATING ABHR STATIONS

Organizations that have protocols requiring the use of an ABHR in presurgical hand antisepsis should install these stations in convenient locations. While such stations should be installed in scrub areas, organizations should consider installing additional units in changing rooms, near bathrooms, and possibly at the entrance to the OR. Although alcohol is a flammable substance, both the National Fire Protection Association (NFPA) and the Joint Commission have reviewed the risks associated with fire and ABHRs and have determined that such risks are negligible. (See Sidebar 6-3 on page 79.)

MONITORING COMPLIANCE EFFORTS

Whatever methods an organization uses to educate the staff on hand antisepsis and encourage compliance with protocols, it is important to monitor compliance over time. There are several ways to do this. Observation is one of the most helpful ways to identify whether perioperative staff behaviors are changing. By posting observers in scrub areas to unobtrusively monitor hand-hygiene practices, organizations can get a sense of whether protocols are being followed. Organizations may want to observe the methods of hand hygiene, the length of time it takes to perform hand antisepsis, and whether re-cleaning is done periodically. These observations can help indicate whether chosen interventions are appropriate and whether further work is necessary. Staff members should be given regular feedback about their hand-hygiene efforts so that continuous improvement can be realized.

Sidebar 6-3. Alcohol-Based Hand Rubs

Some people have expressed concern that ABHRs are flammable. While acknowledging this concern, the Joint Commission believes the typical ABHR dispensers used in the health care setting are of such limited size and volume that their contribution to the development, acceleration, or spread of fire in most situations is small. In a recent survey of 800 facilities reporting a cumulative 1,430 years of ABHR use, no fires attributable to or involving an ABHR dispenser were reported.[7] The NFPA has modified *Life Safety Code (LSC)* requirements to allow for installation of ABHR gel dispensers in egress corridors, subject to certain conditions being met:

- The corridor width is 6 feet or greater, and dispensers are at least 4 feet apart.
- The dispensers are not installed over or directly adjacent to an ignition source, such as an electrical outlet or switch. *Adjacent* is defined as being at least 6 inches from the center of the dispenser to an ignition source.
- In locations with carpeted floor coverings, dispensers installed directly over carpeted surfaces are permitted only in sprinklered smoke compartments.
- Each smoke compartment may contain a maximum aggregate of 10 gallons (37.8 liters) of ABHR gel in dispensers and a maximum of 5 gallons (18.9 liters) in storage.
- The maximum individual dispenser fluid capacity is 0.3 gallons (1.2 liters) for dispensers in rooms, corridors, and areas open to corridors.
- The maximum dispenser size for individual dispensers in areas designated as suites of rooms is 0.5 gallons (2.0 liters).

The Centers for Medicare & Medicaid Services (CMS) is in the process of amending its rules to reflect the NFPA's position.

The situation is a little different with respect to ABHR foam products because all the testing on which the NFPA and CMS decisions are based was done on the gel product, not on foam. However, industry experts have indicated that small-quantity ABHR foam dispensers may be equivalent to those for ABHR gel. Therefore, pending further review, the Joint Commission will allow any ABHR foam installation that meets the location criteria stated previously for ABHR gel.[7]

Note that local or state fire code requirements may differ from the national codes; therefore, an organization should determine and follow the requirements for its particular locale. The best resource for this information is the local fire marshal.

For more information on fire safety, see Appendix 1.

References

1. Semmelweis's revenge: A new era of hand hygiene dawns in health care: ICPs get innovative to monitor compliance. *Hosp Infect Control* 30:113–116, Sep. 2003.

2. Fry D., et al.: Hand hygiene compliance: Step up, reach out. *OR Insider* Nov. 2005, p. 15.

3. Berman M.: One hospital's evaluation of brushless scrubbing. *AORN J* 79:349–358, Feb. 2004.

4. Grabsch E., et al.: In-use efficacy of a chlorhexidine in alcohol surgical rub: A comparative study. *ANZ J Surg* 74:769–772, Apr. 2004.

5. Aksoy A., et al.: An investigation of the factors that affect surgical hand disinfection with polyvidone iodine. *J Hosp Infect* 61:15–19, 2005.

6. Inadequate standards of hygiene in an operating theatre changing room. *The Hospital Infection Society* Sep. 2004, p. 268.

7. Joint Commission on Accreditation of Healthcare Organizations: *FAQs for the 2006 National Patient Safety Goals.* http://www.jointcommission.org/NR/rdonlyres/F758DCC1-D0CA-455B-8031-12AF716660B2/0/06_npsg_faq7.pdf (accessed May 10, 2006.)

Chapter 7: Ensuring Sharps Safety

It Could Have Happened This Way

Dr. Elizabeth Clark, a surgeon at a large metropolitan medical center, was suturing a patient who had just received surgery. She finished suturing the wound and was preparing to tie off the sutures. She was momentarily distracted by a question from a nurse and forgot to remove the needle before tying. As she began tying the sutures, the needle penetrated her finger and drew blood.*

Because of Dr. Clark's sharps injury, the patient was tested for HIV, hepatitis B, and hepatitis C. Blood tests confirmed that the patient was infected with HIV. Dr. Clark followed the hospital's policy on exposure risks and received preventive prophylaxis, regular clinical testing, and psychological follow-up. Six months after the needlestick, Dr. Clark learned that she could finally consider herself safe. But, in the meantime, her mind compulsively played out worst-case scenarios about what would happen to her and her family if she became infected with HIV.

As can be seen in this example, occupational exposure to bloodborne pathogens from needlesticks and other sharps injuries is a serious problem. The Centers for Disease Control and Prevention (CDC) estimates that, each year, hospital-based health care personnel sustain 385,000 needlesticks and other sharps-related injuries.[1] The true magnitude of this problem is difficult to assess because there is believed to be significant underreporting.

* All patient and care-provider names in this book are fictitious.

Although needlestick injuries are the most commonly discussed form of sharps injury, a variety of other sharp instruments can cause serious injury as well, including suture needles, scalpels, and lancets. Such injuries can occur while manipulating a sharp instrument, when handling or passing the device after it has been used, while recapping the instrument, during cleanup following a procedure, or as a result of improper disposal. While most sharps injuries occur among nurses, other staff members can be susceptible, including surgeons, anesthesiologists, surgical technologists, nurse anesthetists, surgery attendants, housekeepers, and students.

The Dangers of Sharps Injuries

Sharps injuries are one of the most common dangers to health care workers, exposing them to many pathogens, including diphtheria, Ebola, herpes, malaria, tuberculosis, and syphilis. Although there are reports of these bacteria, viruses, and fungi transferring through sharps injuries, the most commonly transmitted pathogens are human immunodeficiency virus (HIV), hepatitis B virus (HBV), and hepatitis C virus (HCV.)

The risk of infection from a sharps injury depends on the pathogen involved, the severity of the injury, and the availability and use of appropriate prophylactic treatment.

Strategies to Help Prevent Sharps Injuries

Designing, implementing, and maintaining a sharps injury prevention program requires more than just a few safety devices and one training

Sidebar 7-1. The Needlestick Safety and Prevention Act

In 2000, to draw attention to the issue of needlestick prevention and help organizations implement precautious to address it, Congress passed the Needlestick Safety and Prevention Act. As a consequence of this act, the Occupational Safety and Health Administration (OSHA) revised its bloodborne pathogens standard and requires employers to take three concrete actions to reduce the number of needlestick incidents:

- Review their exposure control plans annually.
- Involve nonmanagerial workers in evaluating selected safety engineered devices.
- Maintain a sharps injury log that ensures employee privacy and contains, at a minimum, the type and brand of device involved in an incident, if known; the location of the incident; and a description of the incident.

Employers are required to abide by OSHA's standards on bloodborne pathogens (see http://www.osha.gov/SLTC/bloodbornepathogens/index.html). Any health care organization that fails to adhere to these standards is subject to a citation from OSHA.

seminar. It requires a coordinated, multidisciplinary effort in which perioperative staff effectively communicate, actively use engineering controls, and engage in safe practices. It requires constant vigilance, regular education, and consistent monitoring. Whenever possible, each organization should craft a training program that illustrates actual exposures within its facility (without identifying the name of the identifier or exposed employee), including (1) type of device, (2) job title of employee, (3) description of event, including location, and (4) suggestions for preventing a similar exposure. Such tailored training reinforces the message that all employees are at risk for such exposures and that prevention is the key.

In case an event occurs, organizations should have plans in place to address the health care needs of the afflicted individual, including his or her emotional as well as clinical needs.

Using a Multidisciplinary Approach to Reducing Sharps Injuries

As with other types of events, the unique operating room (OR) environment affects the nature of sharps injuries. Within the OR, surgeons, scrub nurses, and other surgical team members work in very close proximity to one another, handling the same instruments in a confined space. Staff members can be rushed and inadvertently injure themselves or each other. Consequently, a team approach to preventing sharps injuries is critical if injury rates are to be reduced.[2] Organizations should consider convening a multidisciplinary group to examine and address the issue of sharps injuries. Some actions this group should consider include the following:

- Identify which tasks and procedures are susceptible to sharps injuries and which individuals should receive protective equipment, vaccinations, and education on how to reduce such injuries. The staff should be aware of the requirements associated with OSHA's bloodborne pathogens standards and how the organization complies with those standards.
- Create a written sharps exposure plan, establishing how engineering devices and work practices will be used to prevent sharps injuries. An organization should review this plan each year to make sure it is current. In addition, the organization should keep abreast of new technologies regarding sharps safety and consider adopting these tech nologies, if appropriate.
- Set a policy for postexposure follow-up (see page 86).
- Establish a reporting system for sharps injuries. While OSHA requires a sharps reporting log, an organization may want to consider collecting even more data on sharps incidents. This can help identify trends, patterns, and areas for improvement or further training.

Sidebar 7-2. The CDC's Prevention Workbook

The CDC has created a workbook that is designed to guide health care organizations in designing, implementing, and evaluating a sharps injury prevention program. The workbook contains a comprehensive overview of the literature on the risks and prevention of sharps injuries in health care personnel; guidance on program development; and forms, worksheets, posters, and the like. In addition, it includes the following:

- A worksheet for assessing an organization's sharps injury prevention program
- A tool for conducting a root cause analysis following a sharps injury
- Links to Web sites with relevant information on sharps injury prevention

Organizations can access the workbook at http://www.cdc.gov/sharpssafety/workbook.html.

Using Engineering Controls and Safety Devices

An important aspect of sharps safety involves using safety devices in the OR. These are engineering controls that prevent surgical team members from sticking, stabbing, or cutting themselves with sharps. Following is a discussion of two of the most common safety devices:

- ***Protective scalpels.*** Traditional scalpels can cause more injuries than needlesticks because of their size and sharpness. Consequently, scalpels cause more serious injuries than needles and carry a higher risk of transmitting bloodborne pathogens.[3] Protective scalpels resemble traditional scalpels but have a mechanism that allows the user to either retract the blade into the handle or cover the blade with a protective sheath before and after use. Protective scalpels come in both disposable and reusable varieties. While protective scalpels can help reduce the likelihood of some types of sharps injuries, such as those that occur before use, during transfers, and between uses, they cannot prevent injuries during use. Also, the staff must manually activate the safety mechanisms for a protective scalpel to be effective in preventing sharps injuries.[4]
- ***Blunt-tipped suture needles.*** Suture needles can injure surgical staff members before, during, and after use. The typical suture needle is sharp and can easily pierce the skin. Blunt-tipped suture needles are curved and have a tip that is less sharp than a conventional tapered needle. Although the blunt tip can still be used for suturing, it does not penetrate the skin as easily and is therefore less likely to cause injury. One drawback of blunt suture needles is that they can be slightly more difficult to use than sharp needles, because they require more force to penetrate tissue. While shown to be effective, blunt-tipped suture needles have not gained acceptance among most surgeons. Organizations that choose to use these devices should engage in an education program targeted at gaining approval from the surgical staff. Organizations should identify champions among the surgeons who can influence the acceptance of this equipment.

Although safety devices such as protective scalpels and blunt-tipped suture needles have been shown to reduce the likelihood of sharps injuries, some health care organizations cite the cost of buying these devices as an issue. While cost concerns may be a factor, it is important to note that it is significantly less expensive to prevent a sharps injury than to treat one. Costs of treatment include the direct costs associated with the initial treatment and follow-up treatment of exposed health care personnel, which are estimated to range from $500 to $3,000, depending on the treatment provided.[5] In addition, organizations must consider psychological and emotional costs to the infected worker. This stress can affect performance and attitude and may even cause someone to rethink his or her career choice.

☞ **TIP:** Some surgical team members may resist using new equipment, such as blunt-tipped needles and protective scalpels, and health care organizations may need to confront this resistance to change. By creating a culture of safety in the workplace and identifying champions for change, health care organizations can start to eliminate any opposition to new devices. It can be helpful to include employees in the selection process, as this can go a long way toward facilitating acceptance. ☜

USING SAFE PRACTICES TO REDUCE SHARPS INJURIES

Although safety equipment can go a long way in preventing sharps injuries, such equipment cannot eliminate the possibility of such injuries. Organizations can further reduce the risks associated with sharps by encouraging safe practices among perioperative staff. Some of these practices include the following:

- ***Use a neutral zone.*** A neutral zone is a previously agreed on location on the sterile field where sharps are placed and from which the surgeon, scrub nurse, or surgical technologist can retrieve them without hand-to-hand contact.[8] For example, if a surgeon needs a particular scalpel, the scrub nurse could place the instrument in the neutral zone and remove her hand from the area. The surgeon could then pick up the instrument. This process can reduce the likelihood that the surgical team members could inadvertently stick or cut each other. Organizations should encourage the staff to define such neutral zones before the start of a procedure and engage in hand-free passing whenever possible. The use of a neutral zone is recommended by several professional organizations, including the American College of Surgeons and the Association of periOperative Registered Nurses (AORN).

☞ **TIP:** In addition to using a neutral zone, organizations should consider requiring staff members to verbally announce when they are placing

Sidebar 7-3. Using Technology to Ensure Sharps Safety

As technology advances, so does an organization's ability to prevent sharps injuries. Several technologies virtually eliminate the need to use sharp instruments, such as scalpels or needles. Some of these technologies include the following:

- ***Catheter securement devices.*** Traditionally, catheters have been attached to a patient via sutures or surgical tape. Catheter securement devices do not use sutures and consequently do not require the use of suture needles, thereby reducing the risks of needlesticks.
- ***Laser technologies.*** Some lasers can be used as an alternate means of cutting, rather than a scalpel. Certain electrosurgical equipment can also be used for this purpose. When using laser or electrosurgical equipment, organizations should be aware of the components of the fire triangle and reduce means of fuel, oxidation, and ignition. (See Chapter 4 and Appendix 1.)
- ***Needleless infusion devices.*** The majority of U.S. hospitals have eliminated unnecessary use of needles through the implementation of IV delivery systems that do not require (and in some instances do not permit) needle access. This strategy has largely removed needles attached to IV tubing, such as that for intermittent infusion, and other needles used to connect and access parts of the IV delivery system.[6]
- ***Wound closure technologies.*** Organizations can close small wounds by using surgical glue. This eliminates the need for sutures and suture needles for these kinds of wounds.
- ***Drug delivery systems.*** These systems deliver medication without the use of a needle. One example is a device that delivers insulin to diabetics via a pressurized, microthin stream of medication that penetrates the skin.[7]

Organizations should keep abreast of new technology and consider implementing systems that could prove beneficial to the organization.

items in the neutral zone. This notification can alert other staff members to the presence of a sharp instrument so that those staff members can proceed with caution.

■ ***Practice double-gloving.*** Surgical gloves are a critical barrier between the hands of perioperative staff members and the patient. These gloves can prevent contamination and infection for both the patient and the surgical staff member. While gloves can serve as a barrier, that barrier can be compromised. Perforation rates as high as 61% for thoracic surgeons and 40% for scrub nurses have been reported.[9] The practice of double-gloving involves wearing two pairs of surgical gloves to prevent and mitigate the effects of barrier failure (that is, the ripping, tearing, or puncturing of a glove).[10] On average, an outer operative glove is perforated within 40 minutes of being used. On the other hand, although the puncture of the outer glove is common, corresponding punctures of both the inner and outer gloves are rare. In cases in which a double puncture does occur, the volume of blood is significantly reduced, thus reducing the likelihood of infection.[2]

While double-gloving has been shown to be effective, many staff members have objections to the practice, including poor fit, loss of tactile sensitivity, and increased costs. Following are some suggestions on how organizations can overcome those objections:

■ Work with glove manufacturers to offer a "glove-fitting" station where practitioners can try on many glove options and find a fit that works for them. This will allow practitioners to evaluate performance characteristics of two gloves worn together before entering the OR. This process can be conducted at no cost to the organization.

■ Educate practitioners on the value of double-gloving to justify the time and effort involved in implementation.

■ Enlist the support of individuals who are already double-gloving, such as surgeons or scrub nurses who can testify to the effectiveness of the procedure and allay any concerns.

While some organizations may see double-gloving as an unnecessary increase in cost, the reduction of bloodborne pathogen exposure and the decrease in infections for both patients and providers offer significant savings for health care organizations.

■ ***Regularly change surgical gloves.*** While the use of reinforced gloves or the practice of double-gloving can be helpful in reducing the incidences of sharps injuries and wound contamination, often such injuries are not discovered until after the surgery has been completed. Changing gloves at regular intervals during surgery can reduce the amount of time that a perforated glove is in use. Also, fresh gloves may be less likely to perforate than ones that have been in use for a period of time. In one study, participants changed gloves approximately every 16 minutes, and this was associated with a significant reduction in the incidence of glove perforations for surgeons and nurses.[11]

■ ***Stay out of the way.*** While this may seem like common sense, it is important to note that the staff can reduce the likelihood of experiencing a sharps injury by remaining clear of the suture area while the surgeon is suturing and by remaining clear of the incision area when the surgeon is cutting.[4]

Educating Staff on Sharps Safety

All the preventive measures in the world will not help reduce the number of sharps injuries if staff members are not familiar with the measures and do not use them. For example, safety devices are ineffective if they are not properly activated. Gloves cannot protect people who do not wear them. Neutral zones will not work if perioperative staff forget to use them. The best way to ensure that staff members follow preventive measures is to educate them on those measures and how compliance affects the incidence of sharps injuries. As with other safety education measures, training

Sidebar 7-4. Wearing Appropriate Footwear

Although most discussion about sharps injuries centers on or around the hands and upper body, scalpels, needles, and other sharp instruments can fall to the floor and injure the feet. For this reason, it is important to consider the type of shoes worn in the OR. A recent study by Jerome Barr, M.D., noted that although shoes cannot be relied on to prevent injury, certain shoes are better at preventing injury than others. Rubber shoes and those with an inner lining offer the greatest amount of protection. Steel-toed work boots and some cowboy boots can also be effective.[12]

efforts can take a variety of forms, including in-services, posters, tent cards, role-playing exercises, and so forth.

☞ **TIP:** Some organizations verify the use of safety devices by checking disposal containers to see how many devices have been activated. Depending on the count, the organization may follow up with training or work with champions for change to encourage compliance.[3] ☜

To help with educational efforts, organizations may want to review the literature on sharps injury prevention. The following Web sites can be very helpful in providing current content and teaching tools:

- International Sharps Injury Prevention Society (http://www.isips.org)
- Exposure Prevention Information Network (http://www.healthsystem.virginia.edu/internet/epinet/)
- National Alliance for the Primary Prevention of Sharps Injuries (http://www.nappsi.org)
- The Centers for Disease Control and Prevention (http://www.cdc.gov)
- Occupational Safety and Health Administration (http://www.osha.gov)

Sidebar 7-5. Recommendations for Managing Exposure to HIV, HBV, and HCV

HIV

- Verify the presence of HIV in the patient.
- Administer postexposure prophylaxis, using antiretroviral intervention.
- Provide counseling.
- Document the event and response.
- Test and retest for presence of HIV.

HBV

- Verify the presence of hepatitis B in the patient.
- Verify receipt of hepatitis B vaccine in the exposed individual. Vaccinate if necessary.
- Administer postexposure prophylaxis, using immune globulin or antiviral agents.
- Provide counseling.
- Document the event and response.
- Test and retest for presence of HBV.

HCV

- Verify the presence of hepatitis C in the patient.
- Conduct follow-up hepatitis delta virus testing.
- Provide counseling.
- Document the event and response.
- Test and retest for presence of HCV.

Managing Occupational Exposure

Health care workers who are exposed to sharps injuries face a daunting assortment of short- and long-term consequences. OSHA standards require exposed workers to go through a postexposure prophylaxis procedure. This involves a meeting with a health care practitioner who has experience in sharps injuries. The exposed worker is asked about the injury, what type of device was used, and whether he or she knows the health status of the patient in relation to hepatitis and/or HIV/AIDS. If the patient's health status is unknown, the organization must get the patient's consent for a blood draw.

Not everyone who incurs a needlestick exposure will require management of exposure. Organizations should have policies that describe what types of injuries warrant management. Some things to consider within that policy include the following:

- The type of exposure, such as percutaneous injury, mucous membrane exposure, or non-intact skin exposure
- Type and amount of fluid/tissue, such as blood, fluid containing blood, potential infections fluid or tissue, or direct contact with concentrated virus
- Infectious status of the source, such as presence of HIV, HBV, or HCV
- Susceptibility of the exposed person, such as the individual's HBV vaccination status or his or her HBV, HCV, or HIV immune status[13]

In most cases, infected individuals need not alter their patient care duties. However they should be instructed to follow strict infection control precautions. Staff members should also be advised to refrain from donating blood, plasma, or tissue until after the follow-up period. In addition, they should be advised to use a condom or refrain from having sex until their HIV status is determined.[13]

☞ **Tip:** Anyone exposed to HIV, HBV, or HCV will worry about whether he or she has been infected. Organizations should make sure that any necessary counseling is provided by individuals who are trained to explain the risks and pertinent information to the exposed person in an accurate and compassionate fashion. ☜

A critical piece of occupational exposure management is documentation. As previously mentioned, OSHA requires organizations to keep a log of all sharps injuries. To help collect and document information on a sharps event, some health care organizations create an occupational exposure report form. This form can not only help manage the response to the incident but can also be used to track risky behavior, identify areas for staff education, and reveal systems or processes that require improvement. An example of an occupational exposure report can be found in Figure 7-1 on page 88.

As with any other reporting form or system, organizations must respond to reports of sharps injuries. Trending and benchmarking over time can help identify areas for improvement. Response efforts could include offering further education, implementing revised engineering controls, or further encouraging safe practice patterns.

References

1. Panlilio A.L., et al.: Estimate of the annual number of percutaneous injuries in U.S. healthcare workers [Abstract S-T2-01]. In *Program and Abstracts of the 4th International Conference on Nosocomial and Healthcare-Associated Infections.* Atlanta, Mar. 5–9, 2000, p. 61.
2. Berguer R., et al.: Preventing sharps injuries in the operating room. *J Am Coll Surg* 199:462–467, Sep. 2004.
3. Sharps safety. *OR Manager* 19:12, Jun. 2003.
4. Silverman R.: New guide to safer sharps for the OR. *OR Manager* 20:22–23, May 2004.
5. As quoted at the CDC Web site, http://www.cdc.gov/sharpssafety. (accessed May 5, 2006).
6. Centers for Disease Control and Prevention: *Workbook for Designing, Implementing, and Evaluating a Sharps Injury Prevention Program.* Jul. 28, 2004. http://www.cdc.gov/sharpssafety/workbook.html (accessed May 4,2006).
7. National Alliance for the Primary Prevention of Sharps Injuries: *Primary Prevention Devices.* http://www.nappsi.org/whynappsi6.shtml (accessed May 3, 2006).
8. Stringer B., et al.: Effectiveness of the hands-free technique in reducing operating theatre injuries. *Occup Environ Med* 39:703–707, 2002.
9. Tokars J.I., et al.: Percutaneous injuries during surgical procedures. *JAMA* 267:2899–2904, 1992.
10. LeSac Twomey C.: Double gloving: A risk reduction strategy. *Jt Comm J Qual Saf* 29:369–378, Jul. 2003.
11. Al-Maiya M.: Glove perforations and contamination in primary total hip arthroplasty. *J Bone Joint Surg* 87-B: 556–559, Apr. 2005.
12. Barr J.: Dangers of dermatologic surgery: Protect your feet. *Dermatol Surg* 30:1495–1497, 2004.
13. Berry A.: Needle stick and other safety issues. *Anesthesiology Clin N Am* 22:493–508, 2004.

OCCUPATIONAL EXPOSURE REPORT FORM

Name: ______________________________

Phone number and e-mail address: ______________________________

Date: ______________________________

Time of exposure: ______________________________

Description of the exposure event

Where did it occur?

How did it occur?

What sharp was involved?

How severe was the exposure? Was blood drawn?

Details of the exposure source

Is there known HIV, HBV, or HCV? If so, which one(s)?

What is the status of the patient's disease?

Details of the exposed individual

Has the individual had an HBV vaccination?

Does the individual have HIV, HBV, or HCV?

Details of follow-up

Has the exposed individual received counseling?

Has the exposed individual received postexposure management?

What further work needs to be done?

Who is responsible for managing the follow-up with the individual?

Figure 7-1. *An occupational exposure report form such as this can help organizations comply with OSHA's requirement to keep a log of all sharps events.*

Chapter 8: Preventing Overbooking of ORs

It Could Have Happened This Way

At 3:00 P.M. on a Tuesday, a 13-year-old girl was brought in to the emergency department (ED). She had had abdominal pain for 36 hours and was vomiting. She was febrile and had tenderness on her lower right side. After examining her, Dr. Smith called the operating room (OR) to schedule an emergency appendectomy. Every OR was full with elective cases that were under way. In three of the ORs, surgeons were using their block time. In the fourth OR, a procedure that was scheduled to be finished was running late due to complications. One OR was scheduled to be available at 4:00 P.M., and a further 30 minutes were necessary to clean and prepare that OR for the next surgery. Dr. Smith and his patient had no choice but to wait until the OR was available.*

At 3:50 P.M., the patient's appendix perforated, causing a life-threatening situation. The preoperative process was rushed, and the surgical team was forced to operate under highly stressful circumstances. After surgery, the girl remained in the postanesthesia care unit until an intensive care unit (ICU) bed became free. Because of all the elective surgeries scheduled that day, there was a shortage of beds.

While the surgery was successful, the patient remained in the hospital for three weeks, battling a serious infection. Although the parents were relieved when their daughter made a full recovery, they were frustrated by the situation, feeling it could have been avoided. The entire episode was costly for the hospital both financially and in terms of patient and family satisfaction.

* All patient and care-provider names in this book are fictitious.

In the above example, a different operative schedule could have dramatically changed the outcome. For example, if an OR was available, perhaps the surgery could have been performed before perforation of the appendix. If an ICU bed was available, the patient could have been moved more quickly from the OR and thus have begun the recovery process sooner. In addition, the surgeries performed after this one would not have been delayed.

If an OR schedule is managed well, an organization can reap many benefits, including improved patient care; patient, physician, and staff satisfaction; and optimized revenue. Conversely, if an OR schedule is managed poorly, a patient safety issue can result, and patients, physicians, and staff can become frustrated. In addition, when time in the OR is underutilized, revenues can be lost needlessly.

Creating a schedule for the OR is not an exact science. In fact, it is more of an art, relying on various inputs, factoring in certain circumstances, and depending on historical trends.

Managing the OR schedule should be a team effort. Surgeons, anesthesiologists, OR managers, scheduling staff, and organization leadership should all work together to balance the organization needs with the needs of the patient and the needs of the physicians.

While there are a variety of ways to effectively manage the OR schedule, the following sections offer some strategies commonly used in health care organizations.

Sidebar 8-1. Queuing Theory

Queuing, or waiting line, theory is a mathematical model used to predict flow through a system or process based on historical data. Use of the model can identify bottlenecks and enable reengineering of a process, resulting in reduced wait times and delays and improved efficiencies. While queuing theory has been around for some time, it has only recently been studied in relation to OR scheduling. The Institute for Healthcare Improvement (IHI) has done some important work in this area. More information is available at http://www.ihi.org/IHI/Topics/Flow/PatientFlow/Changes/IndividualChanges/DedicateOperatingRoom(s)forUnscheduledSurgeries.htm.

Managing Block Time Effectively

Using block time is a common practice in most organizations. It can be thought of as a type of reservation; surgeons know that on a particular day at a particular time, there will be an OR reserved for their use. Within that time frame, a surgeon can conduct elective or urgent procedures, several short procedures, or one long procedure. Block time can be assigned to a particular surgeon, practice, or specialty, depending on organization policy. If a surgeon, practice, or specialty regularly underutilizes its block time, it could lose that time, just as a person who does not show up for a restaurant reservation can lose that reservation.

How block time is managed in an organization is a very sensitive issue. Revenue, patient safety, and surgeon and staff satisfaction can all be significantly affected by how an organization manages its block time. For example, if an organization designates block time in all of its ORs, then there may be difficulty scheduling urgent cases. Or, if an organization does not reevaluate the block time of a surgeon who regularly underutilizes it, the organization may not be optimizing its revenues. If surgical staff spend time waiting for surgeons who do not show up for their block time, staff satisfaction issues can result. Following are some strategies organizations should consider for managing block time:

- ***Set clear rules.*** Surgeons should have a complete understanding of an organization's block time policy and what to expect with regard to that policy. They should understand under what conditions their block time could be bumped and how to constructively resolve any conflicts with block time. Rules must be fair and consistent and take into consideration the needs of the surgeon, the organization, and the patients served by the organization.
- ***Consider utilization as one guideline in managing block time.*** Typically, block time is adjusted based on utilization. If a surgeon does not use his or her block time, he or she loses it. While utilization is a typical way of managing block time, it is not the only way or even, necessarily, the best way. Utilization does not take into consideration variable costs, reimbursement issues, or patient waiting times. Some experts feel it is questionable whether utilization can even be estimated precisely.[1] While utilization should be one consideration for managing block time, some other variables to consider include staffing allocations, OR efficiency, predictability of work hours, patient delays, and the balancing of overutilized and underutilized time.[1]

☞ **Tip:** It is unrealistic to think that every surgeon in an organization will use 100% of his or her block time every month. When determining utilization goals, organizations may want to set a more modest expectation, such as an average of 80% utilization over three months.[2] ☜

☞ **Tip:** When measuring utilization of block time, organizations should consider whether they want to use raw or adjusted utilization data. Raw utilization is the total number of hours of cases performed within a block time divided by the hours of allocated block time. Adjusted utilization considers setup and cleanup time within the equation.[3]

Consider the following example:

Surgeon A performs 30-minute procedures that require 15 minutes of setup and 15 minutes of cleanup for each procedure. In an eight-hour block time, this surgeon can complete eight cases. Surgeon B conducts a procedure that is completed in 8 hours. Using raw utilization, Surgeon A would only have 50% utilization of his or her block time, and Surgeon B would have 100%. Using adjusted utilization, both surgeons would have 100% utilization for their block time.[2] ☜

- ***Regularly review block time.*** Organizations may want to consider collecting and reviewing data on block time on a monthly basis. Although changes need not be made every month, monthly data can be reviewed to identify utilization, OR efficiency, staffing allocation, and patient delay patterns that need to be addressed. Organizations that rarely review and tweak the block time schedule may not be aware of the need to revise that schedule and could face significant challenges when it is necessary to make changes. Conversely, organizations that regularly review, adjust, and manage the block time schedule will not meet as much resistance when changes to the schedule must be made. Surgeons should know how block time will be reviewed, what the criteria are for reducing block time, and how frequently changes will be made.
- ***Add an urgent room.*** Many organizations have found it useful to designate an OR for urgent care use. This OR is free from block time and is devoted to unscheduled, urgent, add-on cases. The benefits of this type of room are that it can increase overall efficiency and reduce variability in scheduled surgical flow. Organizations that are regularly bumping other cases for urgent ones should consider this strategy. If an organization has regular gaps in its OR schedules, then such a room may not be necessary. Utilization of an urgent room should never be 100% because that would defeat the purpose of it being available as an urgent room. Although capacity will not be maximized, such a room can streamline the scheduling process and prevent other surgeries from getting bumped. In the long run, an urgent room can save surgeons time, increase their productivity, and improve the bottom line for the organization. For example, St. John's Hospital, an 866-bed, not-for-profit hospital located in Springfield, Massachusetts, designated one of its ORs for add-on and urgent cases. This initiative resulted in a 5.1% increase in patient throughput. In addition, there was a 4.5% decrease in blocked rooms for surgical cases. This number is significant because the block time decreased while the throughput time increased, showing that the add-on room provided the flexibility necessary to get patients in and out more quickly.[4]

☞ **Tip:** If an organization chooses to create an urgent room, it must specifically define what types of procedures can be done in that room. Some definitions could include the following[5]:

- ***Emergent.*** These procedures occur when there is a threat to life or limb and should be performed within 30 minutes.
- ***Priority.*** These cases should be performed within 30 minutes to 4 hours.
- ***Urgent.*** These cases should be performed within 4 to 24 hours.
- ***Nonurgent.*** These cases can be performed after 24 hours.

Cases in the first three categories should be considered for the urgent room.[6] To prevent inconsistency in scheduling, OR schedulers should be educated on what cases can be performed in an urgent room. ☜

- ***Define release times.*** Release times define when the guaranteed availability of a surgeon's block time expires before the day of surgery. These times can be different for different specialties. (See Sidebar 8-2.) For specialties in which cases are booked far in advance, such as joint replacement or cosmetic procedures, release times may be significantly before the surgery date. Conversely, procedures that are

Sidebar 8-2. Possible Release Times for Various Specialties

Cardiac surgery	1 day
General surgery	7 days
Gynecology	7 days
Head and neck	7 days
Neurosurgery	4 days
Ophthalmology	7 days
Joint replacement	14 days
Orthopedics—spine	3 days
Pediatrics	7 days
Cosmetic surgery	14 days
Vascular surgery	2 days
Thoracic	3 days[7]

not booked far in advance, such as cardiac procedures, may have release times that fall 24 hours before surgery. Varying release times can help an organization add more flexibility into the OR schedule. For example, release times that occur far in advance of surgery can be used to free up time to add more urgent cases.[2] If an organization has an urgent room, it should designate a release time for that room. Some organizations keep this schedule open until 24 hours before surgery. Regardless of the amount of time rooms will be kept open, organizations should have a specific policy that is consistently enforced.[8]

Separating Urgent and Elective Cases

Five years ago, if an OR manager was asked to identify the factors that affect the OR schedule, he or she may have given a variety of responses, including tardy surgeons and emergency procedures. Until recently, however, the variability of the elective surgery schedule would not have been mentioned. But a recent study by Eugene Litvak, Ph.D., and his colleagues at Boston University, shows that an elective surgery schedule can stress a hospital even more than the random cases that arrive in the emergency department.[9]

The study shows that during a hospital's busiest time, nearly 70% of diversions from the ICU are associated with variability in the scheduled surgical caseload. In other words, if an organization has an elective surgery schedule that is heavy on some days and light on others, these peaks and valleys in demand for postoperative beds can cause significant patient flow problems.

The study also found that by smoothing out the elective surgery schedule, surgeons could get more work done and increase their revenue.[9]

Consider this example:

The cardiac surgery department of a major medical center has block times on Wednesdays and Thursdays. When the patients operated on during these block times come out of surgery, they go to the ICU. Soon, all the ICU beds are full, and there is no room for patients who come in as emergencies. Consequently, the ED is placed on diversion. Some weeks, if the demand for ICU beds is high enough, some surgical patients are held in the postanesthesia care unit.

To address the issue, the organization reviews its elective surgery schedule. Leadership approaches some surgeons and asks them to switch their elective surgery block times from Wednesday and Thursday to Friday or Monday. The organization also makes ancillary units, such as the physical therapy department, available seven days per week so that any surgery could be scheduled on any day. By eliminating some of the peaks and valleys associated with the elective surgery schedule, the organization provides a more consistent and predictable schedule for the OR. Surgeons' productivity increases, as does their revenue.[10]

Some organizations may hesitate to take on the process of smoothing out the elective surgery schedule for fear of reducing surgeons' productivity or satisfaction, causing surgeons to take their business elsewhere (such as the ambulatory surgery center down the road). However, by using data to examine the effects of such a schedule shift, organizations can anticipate the increase in productivity and revenue that such a shift might cause. By sharing these data and testing schedule changes on a small scale, organizations can earn surgeon acceptance for such a shift.

☞ **TIP:** To help with consistency in OR scheduling, organizations may want to designate particular scheduling staff to the OR. This strategy can help reduce delays in scheduling and minimize cancellations and booking errors. In addition, organizations should consider locating the OR scheduling department within the OR, which may provide schedulers with easy access to the surgical team and allow for more real-time decisions.[12] ☜

SEQUENCING URGENT CASES

Although smoothing out the elective surgery schedule can help improve patient flow through the OR, situations occur in which more than one urgent case arises and a decision must be made as to how to order those cases. While there are many ways to prioritize urgent cases, it can be helpful to take a consistent and systematic approach. One thing to consider when prioritizing urgent cases is each patient's medical deadline. This deadline is the time window for the best possible care, indicating when the case must be started to avoid an increase in morbidity or mortality.[13] For example, the medical deadline for appendicitis in a 13-year-old who has shown symptoms for 36 hours may be 1 hour, as the risk of a perforated appendix is greater among patients having surgery more than 37 hours after the onset of symptoms.[14]

Another thing to consider when prioritizing urgent cases is the estimated case duration. Schedulers can obtain this estimate from surgeons, although these estimates are sometimes inaccurate. Probably the best way to determine case estimates is through historical data on the procedure. Taking an average of all the times such a procedure has been performed can help determine a realistic case estimate.

Sidebar 8-3. Varying Times for Discharge

One way to release ICU beds for incoming surgical patients, and thus prevent backups in the OR, is to spread discharges throughout the day. In many organizations, physicians batch their discharge work, typically to the afternoon. Therefore, patients are not necessarily being discharged when they are clinically ready but rather when the physician has time. This can keep patients in beds longer than necessary. By encouraging physicians to change their rounding habits to spread discharges throughout the day, organizations can free up ICU beds earlier, leaving them available for the patients who need them. Switching rounding habits to visit patients who are being discharged first can free up beds even sooner.[11]

To further prioritize cases, organizations may want to consider the following four ordered priorities[13]:

1. Safety of the patient
2. Surgeon and patient access to the OR
3. Maximizing OR efficiency
4. Minimizing the average patient wait time

The following example describes how one OR manager determined the order of two urgent cases using the previously mentioned criteria:

> On Tuesday at 9:00 A.M., two patients require surgery. Patient A's surgery has a medical deadline of 24 hours and a case duration estimate of 5 hours, based on historical data. Patient B has a medical deadline of 12 hours, and the case duration is expected to be 2 hours. At 1:00 P.M., an OR is about to finish its elective cases. Both surgeons and patients are available. No other OR is likely to finish before 5:00 P.M. Turnover time is 0.5 hours long. Which case should go first?
>
> Looking at this scenario from the perspective of the four priorities, the scheduler determines that the safety of the two patients is not affected

by the sequence because both cases will be done in the same day before their medical deadlines. OR efficiency (expected overutilized OR time on the day of surgery), is also not affected by the sequence. If Patient A goes first, the patient wait time would be 2.75 hours (5 hours + 0.5 hours)/2 patients. If Patient B goes first, the wait time would be 1.25 hours (2 hours + 0.5 hours)/2 patients. Thus, the scheduler schedules Patient B first to minimize the average length of time the patients wait.[13]

Improving On-Time Starts

If a surgery is late in starting, it can throw off the OR for the entire day. Patient wait times can spiral, delays in equipment can result, and staff members can either be rushed to complete a procedure or forced to wait around for the next procedure to start. Patient safety, revenues, and patient and staff satisfaction can all be affected. So, how can organizations ensure an on-time start? There are several actions organizations can take to help procedures begin on time. Following is a discussion of some of them.

ESTIMATING TIME APPROPRIATELY

When scheduling procedures in the OR, organizations must consider how long those procedures typically take to complete. As previously mentioned, OR managers can obtain this information by speaking with the surgeon about the anticipated time of the procedure or by looking at historical records of such procedures. Some organizations use a computer program to help with scheduling. This program takes into account the average time for particular surgeries and, using modeling, determines how best to schedule certain procedures. No matter what method is used, organizations should make sure this method is accurate. Data collection that compares the projected completion time of a particular surgery with its actual completion time can help determine whether estimation methods are accurate.

SCHEDULING SHORT PROCEDURES FIRST

The phrase "short and sweet" has some relevance when it comes to surgery. In general, short procedures have less inherent variability than long procedures and offer greater predictability in determining how long they will take to complete. By scheduling short procedures at the beginning of the day, there is a higher probability that subsequent procedures will start on time or closer to on time than those that follow longer procedures.[15]

STREAMLINING THE PREOPERATIVE PROCESS

The availability of ORs is not the only thing that can cause a procedure to be delayed or cancelled. A preoperative process that does not run smoothly can also derail an OR schedule. A patient safety issue can also arise if all the key information is not available at the appropriate time. For example, if allergy information is not available and a procedure starts without it, medication reactions can occur that cause serious patient harm. Organizations that streamline the preoperative process can help increase the likelihood for on-time starts because all the appropriate information will be available and accessible to the perioperative staff.

Ensuring All Elements of the Universal Protocol. As mentioned in Chapter 2, the Joint Commission requires organizations to follow the Universal Protocol for the prevention of wrong-site, wrong-procedure, and wrong-person surgery. If staff members know when and how to verify the patient, mark the surgical site, and conduct the time-out, and if staff members accomplish these tasks at a consistent time and in a consistent way, then time will not be wasted. How an organization follows the Universal Protocol is, in part, left up to the organization. Organizations that have straightforward, easily followed policies can help streamline the preoperative process and possibly prevent procedures from being delayed. For example, staff members need not spend valuable time searching for information with which to verify the patient, arguing about how to best mark the operative site, or waiting for all team members to participate in the time-out process.

Automating the Preoperative Process. To help streamline the preoperative process, some organizations are turning to technology. Although not required by the Joint Commission, automation can help improve documentation, reduce errors, and save time. For example, organizations can use an electronic medical record that is integrated with a scheduling program to schedule surgery and related appointments, pull lab test results for access in the OR, quickly identify allergies, review medications given, verify medical history, and plan for postoperative care, such as rehabilitation and physical therapy. Some systems can remind clinicians of required steps in the preoperative process. For example, if a nurse is charting in the system preoperatively, a message can pop up to remind him or her to double-check for any medication allergies and verify the surgical site. Reminders can also be set at scheduling to indicate whether a patient requires certain tests, such as an electrocardiogram or a particular blood test. Reminders can be set to various urgencies to highlight issues that must be addressed immediately, such as "the patient is going into surgery in 15 minutes, and the x-rays are not present." The advantage of integration is that once a patient's information is entered into the system, it can be available throughout the surgical episode.[16]

Keeping Physicians on Time

When asked why surgical procedures do not start on time, a nurse may answer, "Because the surgeons are always late." While this may or may not be the primary reason for late starts in an organization, it is one that happens frequently in many organizations. A study conducted in one hospital showed that tardy surgeons caused 60% of delayed minutes.[17] To help improve the timeliness of the start of surgery, organizations should consider improving the promptness of their surgeons. Some ways to do this include the following:

Sidebar 8-4. The Benefits of an Automated Preoperative Process

- More accurate scheduling
- More legible and consistent documentation
- Resource-conflict checking
- Access to real-time patient information in all phases of surgery
- Fewer phone calls for scheduling
- Better workflow management
- Automatic alerts and reminders for clinicians.[16]

- ***Educate surgeons on the importance of being on time.*** While this may seem obvious, it is important to set down expectations regarding on-time surgery starts. Organizations should outline their expectations and the reasons behind them, including improved patient safety, staff satisfaction, and efficient use of resources. In addition, the costs savings of starting procedures on time should be shared.
- ***Sharing start-time data.*** By collecting data on the number of instances in which procedures are late in starting and the reasons for this tardiness, organizations can determine whether surgeon tardiness is a factor. If it is, organizations should consider sharing these data with their surgeons. If possible, an organization should provide each surgeon with his or her on-time record as compared to his or her peers. Making surgeons aware of their relative performance will provide them an opportunity to improve independently, based on factual data.
- ***Reward surgeons that are on time.*** Some organizations reward their surgeons and anesthesiologists for on-time starts. Organizations that track start times of surgeries can determine whether a surgeon or an anesthesiologist has a practice of starting on time. Rewards can take many forms, including financial compensation, first-case time slots, or other incentives.
- ***Set consequences for repeated tardiness.*** Some hospitals enforce a policy that if a surgeon is late multiple

times for surgery, that surgeon loses his or her first-case time slot for a period of time. It is important to note that any penalties for tardiness should be supported by the chief of surgery. The OR manager cannot enforce tardiness policies without the full and complete support of medical staff leadership.

■ ***Gain physician buy-in to scheduling.*** When creating the OR schedule, whether an organization uses block time or not, it is important to involve physicians, including anesthesiologists and surgeons, in the scheduling effort. If the physicians feel they have a say in the process and helped create the schedule, they may be more likely to follow the schedule and start procedures on time when possible.

■ ***Keep the OR manager visible.*** Being visible may be an OR manager's best tactic for getting the first cases of the day started on time. A visible and present OR manager can be available to deal with any patient delays and monitor the tardiness of surgeons. If a surgeon has to pass by an OR manager to get to surgery, it can serve as a reminder to be on time.[18]

■ ***Set the tone up front.*** When negotiating physician contracts, organizations should set the tone up front for on-time starts. If the surgeon or anesthesiologist knows it is a priority for the organization and that tardiness will not be tolerated, he or she will be more likely to comply.

There is no perfect way to create a schedule for the OR. Life in the OR is a balancing act with a series of tradeoffs and compromises. Surgeons, anesthesiologists, OR schedulers, OR managers, and organization leadership must all work together to implement a schedule that is consistent yet flexible, efficient yet malleable, and data-driven yet changeable.

References

1. Dexter F.: What's the best way to adjust block time? *OR Manager* 20:11, Nov. 2004.
2. Patterson P.: A few simple rules for managing block time in the operating room. *OR Manager* 20:1, 9–10, Nov. 2004.
3. Donham R.T., et al.: Procedural times glossary. *Am J Anesthesio* 23:4–12, 1996.
4. Joint Commission Resources: St. John's Regional smoothes OR schedule to ease overcrowding. *Joint Commission: The Source* 3:8–10, Oct. 2005.
5. Questions managers ask on patient flow. *OR Manager* 21: 20, Jan. 2005.
6. Boston hospital sees big impact from smoothing elective schedule. *OR Manager* 2:1, 10–11, Dec. 2004.
7. Mazzei W., Blasco T., as cited in Patterson P: A few simple rules for managing block time in the operating room. *OR Manager* 20:1, 9–10, Nov. 2004.
8. Mazzei W., Blasco T.: What's the best way to manage urgent and emergent OR cases? *OR Manager* 21:1, 10–11, Mar. 2005.
9. Patient-flow fixes ease ED crowding. *OR Manager* 20:5–7, May 2004.
10. Smoothing OR schedule can ease capacity crunches, researchers say. *OR Manager* 19:1, 9–10, Nov. 2003.
11. Managing patient flow: Hospital moves to smooth surgery schedule. *OR Manager* 19:11, Nov. 2003.
12. Marr J.: The positive effects of surgical services redesign. *J Nurs Adm* 35:281–282, Jun. 2005.
13. Dexter F.: How do we sequence urgent cases? *OR Manager* 21:12–16, Mar. 2005.
14. Lau W.Y.: Acute appendicitis in children *Aust N Z J Surg* 57:927–931, 1987.
15. Lebowitz P.: Schedule the short procedure first to improve efficiency. *AORN J* 78:651–659, Oct. 2003.
16. Automation eases the preop process. *OR Manager* 21: 12–14, Feb. 2005.
17. Data helps ASC conquer late case starts. *OR Manager* 21: 25–26, Apr. 2005.
18. Visible manager keeps cases on time. *OR Manager* 21:14, Apr. 2005.

PART IV: Addressing Surgery-Related Risks

Chapter 9: Preventing Anesthesia Awareness

It Could Have Happened This Way

Kimberly Bender was nervous about her cardiac surgery. She was a large woman and was concerned about how her weight would affect her surgery and recovery. She never even considered worrying about her anesthesia.*

After nurses prepped Ms. Bender for surgery, the anesthesiologist began the process of "putting her under." Within a few minutes, the anesthesiologist began speaking to the nurses and surgeon in the room, as if Ms. Bender could not hear him. The surgeons and nurses approached the table, talked for a few minutes, and began the operation. Ms. Bender realized something was wrong because she could still hear the doctors talking. She tried desperately to open her eyes and scream, but she could not. She felt pain during parts of the surgery, including the pull and tug of the scalpel. She felt unable to breathe and wondered if she was going to die.

After the surgery ended and the anesthesia wore off, Ms. Bender told one of the nurses about her experience. The nurse did not seem concerned and told Ms. Bender she must have been dreaming.

But what if she was not dreaming? Anesthesia awareness, also called unintended intraoperative awareness, occurs under general anesthesia when a patient becomes cognizant of some or all events during a surgery or procedure and directly or indirectly recalls those events.

* All patient and care-provider names in this book are fictitious.

The Scope of the Problem

While general anesthesia is an extremely safe process that allows surgeons to perform complex and necessary procedures, it is estimated that between 0.1% and 0.2% of all patients undergoing general anesthesia experience some form of anesthesia awareness. This translates into an average of 30,000 cases per year.

Depending on the type of procedure a patient is having, he or she may be more likely to experience anesthesia awareness. For example, cardiac surgery, obstetric surgery, and severe trauma patients are more likely to experience anesthesia awareness because the level of anesthesia used in these cases is less than in other types of surgery and might not be sufficiently deep to prevent awareness.

Surgeries in which muscle relaxants are used may also be more prone to anesthesia awareness, as the use of muscle relaxants can mask an inadequate amount of anesthesia. Lack of patient movement as a result of the muscle relaxants can also provide a false sense that the patient is appropriately anesthetized when he or she is not.

Awareness can also occur during different parts of surgery. For example, awareness episodes often occur at times of high patient stimulation, such as during intubation or incision. If the delivery of anesthetic drugs is interrupted during surgery, awareness can occur during the interruption.

Different types of patients may be more likely to experience anesthesia awareness. Some patients are more resistant to anesthetic than others. Youth age, female gender, and possibly the chronic use of alcohol and recreational drugs can

Sidebar 9-1. Anesthesia Awareness Is Not Just an Adult Problem

While there have been several studies about anesthesia awareness in adults, there are very few such studies that measure awareness in children. However, one study by Andrew J. Davidson and his colleagues revealed that 0.8% of pediatric cases studied experienced anesthesia awareness. Although the reasons behind anesthesia awareness in children are still unclear, organizations should be aware of the possibility and take actions to address it.[1]

Sidebar 9-2. *Sentinel Event Alert* 32

On October 6, 2004, the Joint Commission issued a *Sentinel Event Alert* on preventing and managing the impact of anesthesia awareness. Anesthesia awareness is generally underrecognized and undertreated, and the *Sentinel Event Alert* was developed to draw attention to the issue. It also offers advice to health care organization leaders on how to prevent anesthesia awareness or, when that is not possible, how to ensure that the needs of patients experiencing anesthesia awareness are appropriately met.

all impact the amount of anesthetic necessary to successfully achieve unconsciousness, and if an inadequate amount is used, anesthesia awareness may result.[2] The incidence of awareness also tends to be increased when the dose of general anesthetic must be decreased and carefully titrated to decrease significant side effects. Consequently, sicker patients undergoing major surgery may be more prone to increased risk of awareness.[3]

Other causes of anesthesia awareness include the misuse or failure of anesthesia equipment, the excessive use of neuromuscular blockers, the use of intravenous (IV) delivery of anesthesia versus inhalation, a lack of vigilance on the part of the anesthesiologist, and an intentional lightening of anesthetic to facilitate a quicker recovery and improve operating room (OR) turnover.

While these are the obvious causes for anesthesia awareness, there are some deeper ones. Lack of knowledge and education about anesthesia awareness is a common problem in many organizations. If staff members are not aware of the issue, they will not make an effort to prevent it. In addition, inadequate patient assessment may cause certain risk factors to be overlooked, and the chance to make appropriate adjustments in the amount of anesthesia delivered may be missed.

The Symptoms of Anesthesia Awareness

Many times it is difficult to identify individuals who are experiencing anesthesia awareness. Typical indicators of awareness include high blood pressure, fast heart rate, hemodynamic changes, or movement. These indicators are often masked by the use of paralytic agents that achieve necessary muscle relaxation during a procedure. Other drugs critical to managing the patient's care, including beta blockers and calcium channel blockers, may also mask awareness.

A patient's experience with anesthesia awareness can take many forms. For example, a patient may hear events during surgery, such as alarms, physicians speaking to one another, or loud music. In addition, he or she could have pain, sensations of weakness or paralysis, and feelings of helplessness, anxiety, panic, and impending death. The patient may be able to mentally comprehend what is happening but most times is physically unable to let anyone know. Sometimes the patient may not remember what happened but starts experiencing painful, often inexplicable, psychological difficulties, including sleep disturbances, dreams, flashbacks, and anxiety. Long-term effects of the experience can include emotional problems, nightmares, anxiety, and even posttraumatic stress disorder (PTSD). Anesthesia awareness can also adversely affect a patient's ability to recover from an operation.

Strategies for Reducing the Risk of Anesthesia Awareness

While preventing anesthesia awareness is an important goal for health care organizations, it is not easy to accomplish. Anesthesia administration requires a delicate balance between too much anesthetic and too little. All patients are different, and determining the quantity of anesthesia appropriate for each patient is not an exact science.

☞ **TIP:** Knowledge is power. Organizations that stay informed about the research and data that explore causes and ways to prevent awareness during surgery will have the power to address this patient safety issue. For example, both the American Society of Anesthesiologists (ASA) and the American Association of Nurse Anesthetists (AANA) have set forth certain recommendations for preventing anesthesia awareness. Incorporating these guidelines[4] into organization processes, as appropriate, can be helpful. The guidelines are as follows:

- Consider premedication with amnesic drugs such as benzodiazepines or scopolamine, particularly when light anesthesia is anticipated.
- Administer more than a "sleep dose" of induction agents if they will be followed immediately by tracheal intubation.
- Avoid muscle paralysis unless absolutely necessary and, even then, avoid total paralysis (by using only the amount clinically required). ☜

To help consistently and appropriately address anesthesia awareness, organizations should develop and implement an anesthesia awareness policy. Staff members should be educated on the policy and understand their role in preventing anesthesia awareness. Some actions the policy should address include the following:

- Identifying patients at risk
- Maintaining equipment
- Monitoring patients for anesthesia awareness
- Reacting to incidents of anesthesia awareness
- Educating staff on anesthesia awareness

Identifying Patients at Risk

As previously mentioned, certain types of patients are at increased risk for anesthesia awareness than others. Identifying these patients before surgery and discussing with them their risk for awareness can help avoid some of the fear associated with the issue. To help identify high-risk patients, the staff should be educated on the characteristics of a high-risk patient and how to identify one. During assessment, nurses should screen patients for a previous history of anesthesia awareness and listen carefully for cues of any risks.

To educate patients on their risks, organizations may want to provide written or verbal information on the topic. The ASA and AANA have created a helpful brochure that can be used for this purpose. Copies of this brochure can be obtained at http://www.asahq.org/patientEducation/Awarenessbrochure.pdf. As with all other forms of patient education, information should be presented in a format that the patient can understand. For patients who are not literate, written material would be inappropriate. Organizations should have a variety of options to address the needs of all their patients.

After providing patient education, organizations might want to obtain informed consent from high-risk patients, so that there is documentation of the conversation between the provider and patient. In addition, organizations should consider interviewing high-risk patients postoperatively to determine whether anesthesia awareness may have occurred.

Maintaining Equipment

Equipment failure is one of the causes of anesthesia awareness that can be virtually eliminated. Regularly scheduled maintenance and testing on anesthesia equipment can help prevent breakdowns that can lead to awareness. In addition, the anesthesia machine and its vaporizers should be meticulously checked before any administration of anesthesia. For more information on anesthesia equipment, see Appendix 1.

MONITORING PATIENTS FOR ANESTHESIA AWARENESS

Monitoring patients for anesthesia awareness can be challenging. Traditional clinical monitoring, such as regularly checking blood pressure and heart rate, cannot reliably identify or prevent awareness instances.[2] As previously mentioned, surgical staff should be alert to patients on beta blockers, calcium channel blockers, and other drugs that can mask physiologic responses to inadequate anesthetics.

There are several ways to monitor patients for anesthesia awareness. A traditional method involves asking patients to respond to questions while they are sedated. Such a question might be "If you can hear me, could you please squeeze my hand, nod your head, or make a noise?" While sometimes effective, there are obvious drawbacks to this method. For example, if a patient is unable to move because of the presence of muscle relaxants or other drugs, then he or she may hear the anesthesiologist but be unable to respond.

Another option is the postoperative interview, which involves asking patients postoperatively whether they experienced any symptoms of anesthesia awareness. Although the interview is important, it will only identify patients after the fact. In addition, it may not capture all patients who have experienced this problem because an individual might have experienced awareness but may not fully remember it or may not be sure enough to report it. Also, a patient may, in fact, remember something that did not occur or perhaps did not occur during surgery. Anesthesia awareness during surgery can be difficult to distinguish from awareness during awakening in the postanesthesia care unit.[5]

New methods to detect anesthesia awareness are being developed, including those that measure brain activity rather than physiological responses during surgery. For example, the bispectral index (BIS) monitor displays a single number derived from bispectral analysis of the patient's electroencephalogram (EEG). This device offers a single "depth of consciousness" figure that perioperative staff can watch like a number on a digital clock. Zero indicates induced coma. A reading of 100 shows that the patient is fully awake. The number 60 reflects the ideal sleep state. Anything higher represents a potentially dangerous state of awareness. Anything lower, and the patient could be oversedated.[6]

The patient state index (PSI) is a measure of hypnosis during anesthesia. Like BIS, it is derived from a patient's EEG. Indices range from 0 to 100, with decreasing values indicating increasing levels of hypnosis.[7] Another monitoring technology is the ocular microtremor. This monitors the "frequency tremor of the eyes caused by extra-ocular muscle activity stimulated by impulses emanating in the brain stem."[8]

Some studies have shown that these monitoring methods can be an indicator of level of consciousness. In fact, PSI has been found to predict consciousness with much better accuracy than hemodynamic variables, such as heart rate and blood pressure.[9] However, some studies say that more research is necessary. Organizations should consider examining these monitoring methods as well as other emerging technologies to see if they are appropriate for the scope of services and mission of their particular organization.

☞ **TIP:** During surgery, conversation should relate to the care of the patient as much as possible. Because patients are most likely to recall emotionally threatening remarks during surgery, it is prudent for the OR staff members to avoid voicing negative or derogatory remarks about patients or their prognoses during surgery, under the assumption that the patients cannot hear them.[4] In fact, some organizations have found it beneficial to voice positive, upbeat statements during surgery to help facilitate recovery. Traffic entering and exiting the OR should be monitored, and conversations and other environmental noises should be reduced where possible.[10]

Reacting to Incidents of Anesthesia Awareness

Sometimes, regardless of an organization's best intentions, a patient can experience anesthesia awareness. In these cases, health care practitioners must be prepared to acknowledge and manage the occurrence. Early identification of the problem is critical to adequately acknowledge the situation. Compassion, respect, and diligence are all important in interacting with a patient who has experienced this problem. Organizations should develop a policy on how the staff should deal with patients who have experienced anesthesia awareness. Components of the policy may include the following:

- Interview the patient after the procedure, taking a detailed account of his or her experience and including it in the patient's chart.
- Apologize to the patient if anesthesia awareness has occurred.
- Assure the patient of the credibility of his or her account and sympathize with the patient's suffering.
- Reassure the patient about what you are doing to help prevent such an event from occurring again. Inform the patient that this event has been documented in his or her medical record and will guide an anesthesiologist in any subsequent surgeries or procedures.[4]
- Explain what happened and why. For example, it may be necessary to administer light anesthesia in the presence of significant cardiovascular instability.
- Offer the patient psychological or psychiatric support, including referral of the patient to a psychiatrist or psychologist.
- Notify the patient's surgeon, nurse, and other key personnel about the incident and the subsequent interview with the patient.
- Follow up with the patient daily during the period of hospitalization and for a period of time after the patient is discharged. This length of time will depend on the patient and his or her particular situation.[10]

Tip: If a patient experiences anesthesia awareness, he or she should be instructed during discharge to communicate this incident to all future anesthesia care providers.[4] Such communication can help the patient prevent this type of incident from happening to him or her again.

Educating Staff on Anesthesia Awareness

While most patients do not experience anesthesia awareness, it does happen often enough that the staff should be aware of the issue and know how to manage patients who experience awareness. As previously mentioned, the staff should be aware of the common causes or risk factors contributing to anesthesia awareness and know how to recognize those risks during assessment. In addition, the staff should know how to monitor patients for awareness during surgery and appropriately respond to any incidents reported after surgery.

Different types of education can be helpful, including in-services, reminders at staff meetings, and articles in newsletters. Patient testimonials may be particularly moving for the staff.

Patients who experience awareness can end up anywhere in the hospital, so educational efforts should be targeted to all staff members, not just individuals in surgical units. If all staff members are aware of the issue, they can work to prevent it and also address it compassionately when it does occur.

Even though one case of anesthesia awareness is too many, the types of anesthesia given today and the critical illnesses of patients will probably never allow the issue to be eliminated completely. However, by making staff and patients more aware of the issue and initiating procedures to prevent anesthesia awareness from occurring, where possible, organizations can reduce the likelihood that a patient will experience this very traumatic event.

References

1. Davidson A., et al.: Awareness during anesthesia in children: A prospective cohort study. *Anesth Analg* 100:653–661, 2005.

2. Myles P.S., Symons J.A., Leslie K.: Anesthetists' attitudes towards awareness and depth of anesthesia monitoring. *Anesthesia* 58:11–16, 2003.

3. Sebel P.S., et al.: The incidence of awareness during anesthesia: A multicenter United States study. *Anesth Analg* 99:833–839, Sep. 2004.

4. Ghoneim M.M.: Awareness during anesthesia. *Anesthesiology* 92(2):597–602, 2000.

5. Eger E., et al.: How likely is awareness during anesthesia? *Anesth Analg* 100:1537–1549, 2005.

6. Planck N.: What's up, Doc? *Time International* 150(35):45, Apr. 27, 1998.

7. Schnieder G.: Detection of awareness in surgical patient with EEG-based indices—bispectral index and patient state index. *Br J Anesth* 91:329–335, 2003.

8. Sigalovsy N.: Awareness under general anesthesia. *AANA J* 71:373–379, Oct. 2003.

9. Kerssens C., et al.: Awareness: Monitoring versus remembering what happened. *Anesthesiology* 99:570–575, Sep. 2003.

10. Buntingame B.: Anesthesia awareness. *AORN J* 81:213, Jan. 2005.

Chapter 10: Preventing Retention of Foreign Bodies After Surgery

It Could Have Happened This Way

Tom Stanton had an abdominal procedure to remove a malignant tumor. For weeks after the procedure, he experienced tremendous pain and ran a slight fever. He was unable to sleep, eat well, or go to work. He was understandably concerned about why he was in so much pain. After several weeks with no improvement, Mr. Stanton visited his physician, complaining of severe discomfort, and the physician ordered a computer topography scan. This revealed a 13-inch malleable retractor that had been left inside Mr. Stanton's body after surgery. Mr. Stanton had a procedure to remove the retractor and fully recovered. He was relieved that his pain had subsided but was frustrated that he had had to experience it in the first place. While his body was reacting to the foreign object, it was unable to fully address his cancer. Mr. Stanton wondered what long-term effects of this episode might occur.*

Stories such as this one are not uncommon in the health care field. In fact, a recent study in the *New England Journal of Medicine* estimates that surgery tools and supplies are unintentionally left in more than 1,500 people each year.[1] Because this estimate is based on malpractice claims, study authors suggest that the actual rate of occurrence is probably much higher.

The Consequences of Leaving Behind Foreign Bodies

In many cases in which a foreign body is unintentionally left behind, the offending item is removed and the patient recovers; however, unintentionally retained foreign bodies can lead to severe illness and sometimes death. Mortality rates resulting from retained items have ranged from 11% to 35%.[2] Retained foreign bodies can also cause bowel perforations, organ damage, sepsis, and acute pain.

From the organization's perspective, retained foreign bodies can lead to lawsuits, financial losses, and disciplinary actions, not to mention the time and resources required to remove the foreign body and address any possible infections that may have resulted from the incident.

Risk Factors for Retaining Foreign Bodies

Historically, when a foreign object was left behind during surgery, the surgeon performing the procedure would be accused of negligence. But, as with other types of health care errors, the reasons behind this issue go far beyond that. Following is a discussion about some reasons things get left behind.

Counting Policies Are Not Followed Adequately

Counting policies that require the staff to count equipment before, during, and after surgery have long been an effective way to prevent the retention of foreign bodies. However, these policies are effective only if they are followed, and even then, sometimes an object is overlooked. The previously mentioned study shows that in nearly 90% of cases of unintentionally retained foreign objects, a count was performed and

* All patient and care-provider names in this book are fictitious.

Sidebar 10-1. Making It "Reviewable"

To help organizations address the issue of retaining foreign objects after surgery, the Joint Commission now includes this issue in its list of reviewable sentinel events. The hope is that organizations can become more familiar with the risk factors associated with this issue and learn from other organizations how to address them.

showed that all equipment was accounted for.[1] Obviously, in these cases, the count was incorrect. So how did that happen?

Counting errors can occur when unpacking items during surgery preparation or when reconciling items at the end of the procedure. The counting procedure may be influenced by human factors, such as fatigue, distractions, stress, and interruptions. The staff may lose count or be rushed to finish. Counts can be interrupted by pages and people talking. A nurse may be pulled of the task of counting and forget to go back and complete the count. Concurrent counting by two individuals may not work as intended if one person is distracted. The sequence of the count may not be followed and, depending on the organization's policy, certain items may not be counted.

Certain Procedures Present More Risk

Certain procedures have an increased risk of resulting in retention of foreign bodies, including the following:

- Emergency surgeries
- Complex procedures
- Surgeries that include unplanned procedures, such as an appendectomy that turns into an operation for colon cancer. Such a switch requires more instruments and supplies and may require a change in surgical staff.
- Surgeries that take place on off hours, such as nights and weekends
- Cases in which sponges are used as packing and/or for retraction
- Surgeries involving patients with a high body-mass index[1,3]

During many of these procedures, staff members can be rushed or distracted, and they may need to introduce new equipment. In these cases, the counting process can get quite confusing and complicated. "There is a high risk for retaining foreign bodes any time there is a change in surgical teams during a procedure or any time the pace and speed of the procedure is fast. "These situations create opportunities for breakdowns in communication," says Peter Angood, M.D., vice president and chief patient safety officer for the Joint Commission.

Some less obvious risk factors for this type of event include ineffective communication between surgical team members, worker fatigue, staff turnover, level of surgeon experience, and changeovers of house staff in teaching hospitals.[4]

Strategies for Preventing the Retention of Foreign Bodies

Retaining objects in a patient after surgery is considered to be a preventable occurrence and can be virtually eliminated if organizations take a systematic and proactive approach to the issue. The following sections offer some strategies organizations should consider.

Reviewing the Counting Policy

An organization's counting policy should be standardized for all procedures and be clear and concise. An effective policy should outline what needs to be counted, when it needs to be counted, how it should be counted, and who should do the counting. Organizations should consider convening a multidisciplinary group to conduct a review of the current counting policy and determine what, if any, revisions are necessary. Team members could include representatives from nursing and surgical staff, the biomedical department, organization leadership, and purchasing.

Determining What to Count. Ideally, perioperative staff should count all items that enter the sterile field. However, depending on the procedure, this may not be practical or cost-effective. Organizations should work with the staff to determine what must be counted for each different type of procedure. This may include sponges, needles, other sharps, and bags. It is recommended that sponges be counted in every operation, including labor and delivery.[1] Instruments, such as forceps and retractors, should also be counted for all procedures in which there is a likelihood that an instrument could be retained. Retention of surgical instruments accounts for approximately one-third of retained item case reports.[5]

☞ **Tip:** One of the risk factors for retaining foreign bodies after surgery is the presence of too many supplies. Excess supplies can lead to potentially complex and time-consuming counting procedures. To address this risk, organizations may want to consider refining instrument sets so as to reduce the number of supplies and make counting more efficient.[6] ☜

Determining When to Count. The appropriate time to count items varies depending on the type of procedure and the type of equipment. In the case of sponges and sharps, which are commonly retained foreign objects, the Association of periOperative Registered Nurses (AORN) recommends that they be counted at all the following times:

- Before the procedure to establish a baseline
- Before closure of a cavity within a cavity
- Before wound closure begins
- At skin closure or end of procedure
- At the time of permanent relief of either the scrub person or the circulating nurse

Determining How to Count. To help minimize confusion and miscounts, organizations should consider requiring a consistent order for counting. For example, the staff could start counting items in the surgical field and move out. A scripted order may be helpful. The staff should not disperse instruments to another area of the surgical field, such as a Mayo tray, until an initial count is complete. The risk of forgetting to count an item increases when an instrument is put in two locations.[2]

Sidebar 10-2. Where Are Foreign Bodies Being Left?

Gawande et al.'s study found that foreign bodies were left in some locations more often than others[1]:

Abdomen or pelvis	54.0%
Vagina	29.0%
Thorax	7.4%
Other locations (including the spinal canal, face, brain, and extremities)	17.0%

AORN recommends that sponges be separated, counted out loud, and viewed concurrently during the count procedure by two individuals, one of whom should be a registered nurse circulator. Sponges should also be separated so the numbers of sponges as well as their radiopaque indicators are clearly visible. All components of the sponge should be accounted for.[7] If a sponge is soiled, it should be removed from the surgical field and placed in a previously identified location, such as an impervious drape on the floor. If a sponge is added to the surgical field, it should be counted and recorded.

Documenting the Count. Surgical teams should consistently document all sponge, sharp, and instrument counts. Some organizations use a whiteboard located in the operating room (OR) as the primary count recorder. Counts should also be recorded in the patient's intraoperative record. According to AORN, this documentation should include at least the following:

- Types and number of counts
- Names and titles of personnel performing the counts

- Results of the counts
- Notification of the surgeon regarding any count discrepancies and other actions taken
- Instruments intentionally retained within the patient, such as sponges intentionally left as packing
- If necessary, reasons counts were not performed or completed per organization policy[5]

Individuals should only document what they count. If someone documents for someone else, the risk of double-counting can be present.

☞ **TIP:** To assist with documentation, an organization might want to create a preprinted sheet that lists the contents of surgical trays and any necessary instruments for a particular procedure. The staff can use the sheet as a checklist for packing, unpacking, and before closure. Each time a count is done, it can be documented in a column on the sheet. The staff can easily see when the count is off because the columns will not match. These count sheets should group like instruments, such as retractors or needle holders, under their specific names. Instrument names should be standardized and consistent, and those names should be used on any count sheets.[8] ☜

Determining Who Should Count. Counting should be done by more than one person at the same time. As previously mentioned, AORN recommends that sponges, sharps, and instruments be counted audibly and viewed concurrently by two individuals, one of whom should be a registered nurse. Interruptions should be minimized so as not to distract the individuals who are counting. The surgeon should verbally acknowledge the count and should never be allowed to waive the counting process.

EDUCATING THE STAFF

Education efforts should discuss the importance of preventing the retention of foreign bodies and describe the consequences if an object is retained. Case studies can be particularly helpful in communicating the seriousness of the issue. Education efforts should also familiarize the staff with the policies relating to the retention of foreign bodies and outline the ways compliance and competency will be measured. The staff should know for a particular procedure what must be counted, when, how, and by whom. Role-playing exercises, demonstrations, and in-services may be helpful in this effort.

ENSURING EFFECTIVE COMMUNICATION BETWEEN TEAM MEMBERS

In surgical teams that communicate effectively, any member of the team feels comfortable raising concerns about a count discrepancy. The organizational culture should allow any surgical staff member to stop the procedure if the count is off.

Preventing the retention of foreign bodies should be a multidisciplinary effort, and responsibility should not rest only on the nursing staff. While nurses may be responsible for the counting, the entire surgical team should be aware of the count. If there is a discrepancy, it should be up to the entire surgical team to resolve it.

PERFORMING X-RAYS ON ALL HIGH-RISK PATIENTS

To help verify the count in high-risk patients, organizations should consider establishing a policy that requires all high-risk patients to receive an x-ray to verify the absence of foreign bodies. Some high-risk patients to consider x-raying include the following:

- Patients with high body mass index
- Cases that involve multiples surgeons or multiple surgical fields, such as abdominal or thoracic surgeries or procedures that run particularly long
- Emergency cases in which the pace of the surgery is fast, such as ruptured aortic aneurysms or emergency C-sections[9]

"While the practice of obtaining x-rays for high-risk patients is important, it is not always foolproof. The staff must review the x-rays prior to the patient leaving the surgical suite to ensure that the x-ray has captured all the anatomic areas that were operated on," says Angood. "An x-ray is not effective if it misses areas in which equipment may be 'hiding.'"

X-rays can also produce false positives. One study conducted by Kaiser et al. showed that in 3 of 29 cases in which intraoperative x-rays were used to detect sponges, the x-ray was falsely negative.[10]

To effectively use x-rays in capturing retained equipment, organizations must use equipment and supplies with radiopaque markers that will be visible during x-ray. An x-ray cannot capture retained equipment without such markers. In addition, size can affect the ability to capture foreign objects on an x-ray. Smaller needles, scalpels, and other instruments may not be clearly visible.

Minimizing the Rush

One of the risk factors for retaining objects after surgery is staff rushing. While sometimes the speed of a procedure cannot be helped–such as during an emergency–organizations can help minimize this risk factor by smoothing the OR schedule (see Chapter 8), avoiding back-to-back surgeries if possible, appropriately staffing the OR, effectively coordinating vacation schedules, and reducing staff turnover.

Minimizing Distractions

As previously mentioned, pages, conversations, and nurses leaving the room to retrieve equipment can all distract surgical staff members from the counting process. Organizations should consider working to minimize these disruptions. For example, organizations may want to minimize paging in the OR and allow only minimal communication during a patient count (see pages 12–13), so the person counting does not get distracted. In addition, when extra supplies or trays are needed, circulating nurses should be able to call out for them instead of leaving the room during the count.[9]

Monitoring Success

Organizations should continuously review compliance efforts with counting procedures. As mentioned in previous chapters, the only way to determine whether staff members are following policies, procedures, and protocols is to measure their compliance. There are many ways to measure compliance including observation and medical record audits. Organizations that use a checklist to help the staff manage counting procedures can use that checklist as an auditing tool. As with other monitoring efforts, consistent feedback should be given to staff members on how they are doing regarding compliance efforts, areas of success, and areas that need to be improved.

Responding to a Count Discrepancy

If there is a discrepancy in a count, the surgical team should take steps to locate the missing item. Following are several steps organizations should consider for use in a protocol for responding to a count discrepancy:

1. Report the discrepancy to the surgeon. This is typically done by the circulating nurse, and the surgeon should verbally acknowledge the count discrepancy.

2. Suspend the procedure if the patient's condition permits.

3. Inspect the operative site.

4. Visually inspect the area surrounding the surgical field, including the floor, kick buckets, and linen and trash containers. No trash or linens should be removed until the patient leaves the room.

5. If the patient's condition permits, take an intraoperative x-ray and read it before the patient leaves the OR.

6. Document all measures taken on the patient's record.

7. Report the incident, using the organization's reporting system.

8. Review the incident for cause, effect, and prevention.[5]

As with counting policies and other initiatives to reduce the retention of foreign bodies, organizations should solicit multidisciplinary involvement in developing protocols that outline the proper response to a count discrepancy.

Electronic and Radio-Frequency Identification Tagging

To help prevent the retention of foreign bodies, some organizations are investigating emerging technology that involves electronic or radio-frequency identification (RFID) tagging. With this technology, supplies and equipment are marked with electronic or radio-frequency markers, which can be detected by a wand that passes over the patient or by passing the patient under an RFID reader. This type of technology is used every day by commercial couriers who are able to deliver millions of packages a day at a low cost and track all of them at the same time.

Organizations should watch for this type of technology as, in the future, it may become an efficient way to prevent the retention of foreign objects. Angood warns that this new technology is not a panacea. It may make tracking equipment easier; however, organizations still need a culture that promotes effective communication and procedures that prevent human error to eliminate the retention of foreign bodies after surgery.

References

1. Gawande A., et al.: Risk factors for retained instruments and sponges after surgery. *N Engl J Med* 348:229–235, Jan. 2003.

2. Porteous J.: Surgical counts can be a risky business! *Can Oper Room Nurs J* Dec. 2004.

3. Study offers evidence on items left behind during surgical cases *OR Manager* 19:1,10–11, Mar. 2003.

4. McLeod R., et al.: Risk factors for retained foreign bodies after surgery. *Can J Surg* 47:57–58, Feb. 2004.

5. AORN's recommended practices for sponge, sharp, and instrument counts is up for review. *Infection Control Today* Apr. 1, 2005. http://www.infectioncontroltoday.com/hotnews/54h1945629067.html?wts=20060509125416&hc=731&req=AORN's+and+recommended+and+practices (accessed May 10, 2006).

6. Projects for improving safety of counts in the OR. *OR Manager* 19:11–12, Mar. 2003.

7. Farrell V.: Auditing and improving sponge count process. *OR Manager* 19:13–14, Mar. 2003.

8. Erickson S.: Left behind: Counting surgical instruments reduces medical errors. *Material Managing in Healthcare* Jul. 2003, pp. 32–34.

9. Minnesota takes extra steps on wrong surgery, retained items. *OR Manager* 21:1,18–20, Apr. 2005.

10. Kaiser C., et al.: The retained surgical sponge. *Ann Surg* 244:79–84, 1996.

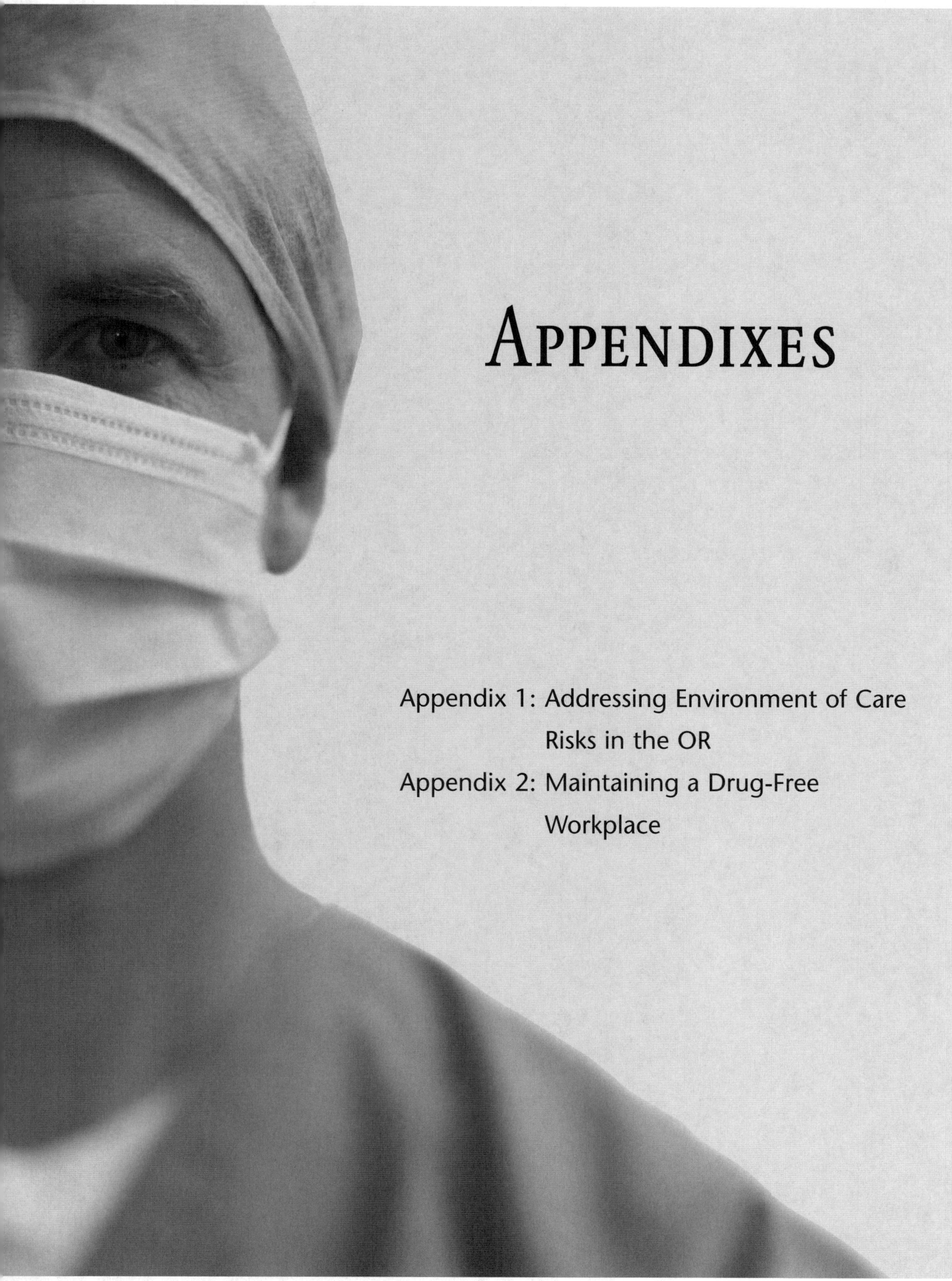

Appendixes

Appendix 1: Addressing Environment of Care Risks in the OR

Appendix 2: Maintaining a Drug-Free Workplace

Appendix 1: Addressing Environment of Care Risks in the OR

No discussion of operating room (OR) risks would be complete without an examination of the environment of care (EC) issues associated with the OR. Utilities management, hazardous materials management, security, and safety all play roles in the safe functioning of the perioperative environment. This appendix provides a brief look at some of these issues and offers suggestions on how to address risks in the OR environment.

Ensuring Awareness of Scheduled Drills

In the middle of surgery, a fire alarm sounds. There is no fire in the OR, but the alarm goes off, alerting the staff to a potential fire in the building. Does the perioperative team stop, stabilize the patient, and evacuate? Or does it keep working and ignore the alarm? The answer to these questions depends largely on whether the fire alarm is a drill or a real fire emergency.

When scheduling a fire drill, organizations should consider the impact to the OR and schedule the time of the drill to cause the least amount of disturbance. For example, the perioperative team should know when the drill will occur and be prepared to ignore it. To prevent distractions, alarms should not sound in the OR. If this is not possible, the OR schedule should be reviewed so that a procedure requiring intense concentration and focus is not under way during a drill.

It may be a good idea to practice a worst-case-scenario fire drill, involving an OR patient in the middle of surgery who must be evacuated; however, such a drill should not involve an actual patient. Scenario drills involving role-playing may be helpful to determine the mechanics of evacuating a patient in the middle of major surgery.

Standard EC.5.30. The organization conducts fire drills regularly.

In addition to coordinating fire drills, organizations should plan for any scheduled utility shutdowns, such as power or medical gas disruptions. When scheduling a shutdown, consider all the ORs that could be affected and plan accordingly. In this case, it is best to be overly cautious. Consider the following scenario:

> A small internal renovation project at a community medical center necessitated the relocation of the medical gas control panel for a patient-care floor. At approximately 4 A.M., hospital operations personnel conducted a scheduled 15-minute interruption of supplemental oxygen lines to the east wing in preparation for the renovations. Nursing staff in the east wing were notified of the scheduled shutoff. To ensure the safety of the patients on the east wing, each patient using piped oxygen was transferred to a portable oxygen unit.
>
> As scheduled, the lead mechanic shut off the oxygen supply valve, unknowingly interrupting piped oxygen supplies to the west wing as well as the east wing. An emergency alarm system alerted appropriate personnel in the west wing. Nurses and respiratory therapists on several floors of the west wing immediately responded by using the existing portable oxygen tanks on

the nursing units to support the patients in greatest need of supplemental oxygen and by assessing other patients. Unit personnel immediately notified the physical plant staff to take action to restore oxygen supplies. However, one patient died 10 days later as a result of the event.

To prevent such occurrences, EC professionals must carefully plan any scheduled shutdowns and consider all possible areas that could be affected. It may be helpful to use failure modes and effects analysis (FMEA) or a similar proactive risk-assessment tool to anticipate any problems with a shutdown and address those issues before the shutdown occurs. (For information on using FMEA to proactively identify risks, see the Joint Commission publication *Failure Mode and Effects Analysis (FMEA): Proactive Risk Reduction,* 2nd edition.)

Other suggestions for effectively communicating about utility shutdowns include the following:

- Process utility shutdown requests in writing, to ensure that the verification of actual utility distribution, concurrence by clinical staff, and administrative approval are documented.
- Locate the physical plant manager and the director of patient care services or OR manager on site and agree to proceed before implementing a planned utility shutdown.
- Use technology to prevent problems. One organization installed a computerized design program in its facility, which allows utility drawings to be updated as repairs, additions, and changes are made so that all building utility blueprints accurately reflect the existing utility systems.[1]
- Engage in a final communication between all affected areas of the organization and the supervisor in charge of the utility shutdown just prior to the shutdown. This communication should include verifying that all backup personnel who are providing portable/emergency utility support (such as tanked gases and battery-powered equipment) are ready.
- Do not make repairs or alterations to the utility system after the initial shutdown until it is ascertained that no unanticipated area of the organization is affected.
- Maintain continual contact between the supervisor in charge of the shutdown and the patient care supervisor, as well as areas affected by the utility shutdown, to monitor for any unanticipated occurrences.
- Do not release supplemental supplies and additional staff until the utility system is verified to be back in working order.

Managing Hazardous Materials and Waste

The operating room is one of many locations in a health care organization where hazardous materials and waste can be found. Organization staff members should be aware of the types of hazardous materials with which they work and be trained on how to properly handle and dispose of those materials. Education efforts for perioperative staff members should address all aspects of hazardous materials and waste, including the following:

- Spill response and cleanup
- Material safety data sheets (MSDS) for hazardous products. Staff members should be familiar with the MSDS for the materials with which they work.
- Proper handling, storage, transport, use, and disposal of hazardous materials and waste
- Proper personal protective equipment (PPE)

Training should be based on the most current information about the products. The information used to support the hazardous materials and waste program should come from several sources and be gathered on an ongoing basis. Primary data sources could include the following:

- Manufacturers
- Suppliers
- Regulatory agencies

- Environmental tours
- The materials management department

Other sources of information may include appropriate safety references or data pertaining to biohazardous materials obtained from the Centers for Disease Control and Prevention (CDC) or from the organization's own infection control and safety programs. The constant changes and multitude of new products classified as hazardous make it imperative that organizations be aggressive in obtaining and maintaining current information and in passing it along to employees.

Standard EC.3.10. The organization manages its hazardous materials and waste risks.

☞ **TIP:** Certain hazardous chemicals should not be allowed in the OR. For example, glutaraldehyde is a hazardous material that, if spilled during a procedure, can cause significant problems. Because a patient may not be movable during surgery, the spill cleanup would have to wait until after the surgery completed, thus exposing the surgical team and patient to the hazardous chemical. For this reason, materials such as glutaraldehyde should not be allowed in any patient care area, including the OR. ☜

DISPOSING OF WASTE

Cleaning up in the OR can be a dangerous undertaking. Infected materials, sharps, and solid waste can all present hazards to the OR environment and the surgical staff. Regulated medical waste is waste that requires special handling and could include the following:

- Contaminated items that would release blood or other potentially infectious materials in a liquid or semi-liquid state if compressed.
- Items that are caked with dried blood or other potentially infectious materials and are capable of releasing those materials during handling
- Contaminated sharps
- Pathological and microbiological wastes containing blood or other potentially infectious materials[2]

Although different organizations have different definitions of what constitutes regulated medical waste, each surgical staff should be aware of its own organization's policy and follow it accordingly.

Within the OR, regulated medical waste, sharps, recyclables, and solid waste should each have separate containers that are clearly labeled and easily accessible.[3] Staff should not have to travel far to dispose of waste, and the process for disposal should be clear.

According to the Association of periOperative Registered Nurses (AORN), all regulated waste, including liquid or semi-liquid blood or other potentially infectious materials, should be placed in red bags.[4] All red bags should be transported to a secured area in the facility designated for hazardous waste. The bags should be placed in this area for storage until it is time for transportation to the waste-management processing plant or facility. Hazardous waste must be packaged according to guidelines published by the U.S. Department of Transportation so that it may be transported safely. Transportation guidelines require the following of packages or containers:

- Be break resistant
- Be impervious to moisture
- Be leak resistant
- Be puncture resistant for sharps
- Be rigid
- Be sealed to prevent leakage during transport
- Be tightly lidded or stoppered for fluids in quantities greater than 20 mL
- Have sufficient strength to prevent tearing or bursting under normal conditions of use and handling
- Weigh 882 lbs (400 kg) or less for nonbulk maximum net mass or 119 gal (450 L) or less for nonbulk maximum capacity[5–7]

Each community has regulations for the disposal of regulated medical waste. Organizations should check with local authorities to learn the correct methods for disposal of regulated waste in their communities.

Documentation of the collection and handling of these materials can be maintained to allow continuous monitoring and evaluation of the program and, in some cases, to meet legal requirements.

Addressing Security

Some of the security risks commonly found in a hospital or an ambulatory care facility, such as theft of personal property or assault, are not as common within the OR. In some ways, patients and staff are relatively protected in this environment. However, there are some risks of which staff members need to be aware:

- ***Personal conflict.*** When highly trained professionals work in close proximity under stressful circumstances, personal conflicts can arise. Add someone with a bad temper to this mix, and potentially violent situations can erupt. To address this possible risk, organizations should offer education and training to perioperative staff on how to positively resolve conflicts, communicate effectively, and operate as a team. (See Chapter 1.)
- ***Rape.*** Rape in the OR is highly unlikely. However, within the postoperative area, a patient is often alone and not fully aware of what is going on around him or her. He or she may be vulnerable to a variety of negative situations, including rape. Organizations should have in place systems to monitor patients in the postoperative area and limit access to approved personnel only.
- ***Loss of property.*** Mr. Smith didn't want to take off his wedding ring, and the preoperative nurse forgot to ask him to do so. Consequently, surgery was about to start, and Mr. Smith was still wearing his ring. The nurse asked Mr. Smith to take the ring off and assured him she would keep it safe in her pocket until after surgery. The procedure took five hours, and the nurse completely forgot about the ring. At the close of the procedure, the nurse took off her gown and threw it, along with Mr. Smith's ring, in the laundry receptacle. Although such oversights do not happen often, organizations should have a formal process for collecting and safely storing jewelry before a procedure. For example, jewelry may be collected in a plastic bag with an ID tag and stored in a secure place.

☞ **Tip:** Because security is not often an issue in the OR, many security professionals may not be familiar with the OR setting. Organizations should consider allowing security personnel to tour the OR and gain familiarity with it. Security professionals who will respond to incidents in the OR should know the proper infection control precautions to take, such as wearing PPE and regularly disinfecting hands. ☜

Standard EC.2.10. The organization identifies and manages its security risks.

Security During Off Hours

Most surgery is not conducted at night or on the weekends, so procedure rooms are empty during off hours. ORs can store a variety of equipment and instruments that, if used incorrectly, could harm patients or staff. Organizations should make sure all ORs are effectively secured when not in use. In addition, access to these rooms after hours should be strictly controlled. For example, organizations may want to use a card access system to monitor who comes into and goes out of the OR. This system could also monitor the frequency and duration of OR visits. Only designated and approved individuals should be granted card access to the OR. In some cases, organizations may want make available only a few cards that must be checked out from and returned to the security department.

CLEARING VISITORS OR OBSERVERS

During some OR procedures, there may be more than just the patient and the surgical team present. Vendor representatives, students, and other observers may be present as well. If more than just the surgical team will be present, any observers should receive the appropriate vaccinations, safety orientation, and criminal background checks. Patients should give permission to have visitors and observers in the OR. Because different types of observers and visitors require different safety precautions, organizations may want to develop a grid that outlines what information, vaccinations, or approvals are necessary for each type of observer. This grid can also serve as a checklist to ensure that proper information and approvals are gained before the start of surgery.

☞ **TIP:** It is important to consider the patient's privacy when inviting students or other visitors into the OR. For example, students and any other observers should not have access to the surgery schedule. Although it is important to provide learning experiences for students, surgery schedules contain confidential patient information that must be carefully protected. According to AORN, the surgery schedule should be shared only with authorized individuals after all patient identifiers have been removed. As it may be possible to identify patients without knowing their actual names, schedules that contain any information that could be used to identify specific patients should not be distributed randomly or given to individuals who do not understand the importance of securing the information. In addition, schedules should not be placed in areas where unauthorized individuals can view the information, such as at unattended volunteer desks, non-staff member areas, or admitting desks.[6] ☜

Reducing Noise

Machines are humming, people are talking, rock music is playing loudly, and the overhead patient system keeps asking for Dr. Miller to please call the operator. Such an environment is enough to give someone a headache, and it does. Noise in the OR can be, at the very least, irritating and, at worst, a patient safety hazard.

Consider the following incident reported to the Institute for Safe Medication Practices[7]:

> During a carotid endarterectomy, the surgeon verbally ordered a dose of heparin. Both the anesthesiologist and anesthesia resident heard the surgeon order 2,000 units of heparin and, therefore, 2,000 units were administered. At an appropriate interval after administration, the patient's activated clotting time was checked and found to be 201 seconds (control 145 seconds). The surgeon's desired level was 300 seconds to 350 seconds, so another 1,500 units of heparin were administered. After waiting the appropriate length of time, the anesthesiologist checked the clotting time and determined that the patient's level had dropped to below 200. When the surgeon questioned why the desired level was not being achieved, the anesthesiologist reviewed with the surgeon the doses that had been given. At that point, the anesthesiologist learned that the original order from the surgeon was for 10,000 units, not 2,000 units. Both the anesthesiologist and anesthesia resident had heard the surgeon incorrectly.

This example illustrates how important it is to be able to hear in the OR. Reduced noise levels can help people hear better and, consequently, perform their jobs better.

To help reduce noise levels in the OR, organizations should consider the following strategies:

- Develop a policy that minimizes or eliminates overhead paging in the OR.
- Examine medical equipment for noise risks. Clinical alarms, medical carts, and other equipment can all make noise. By examining the equipment in the OR and determining whether and how the noise of that equipment can be reduced

Sidebar A-1. The Hazards of Playing Music in the OR

There are several issues associated with playing music in the OR. One problem is that it is difficult to find music that is acceptable to all personnel involved in a procedure. Not everyone likes the same kind of music, and what may be energizing, soothing, or otherwise pleasant to one person may be irritating to another. No matter who chooses the music, it is likely to be stressful or annoying to one or more of the other providers in the surgical theater. The fact is that few people do their best work when stressed or annoyed.

Perhaps the greater problem with music in the OR is the decibel level at which it is played. Adding the noise of a loud radio to the existing noise of essential, procedure-related equipment, such as monitors, alarms, or suctions, is not in the patient's best interest.

As previously mentioned, verbal orders are difficult to hear in the OR because surgical team members usually cover their mouths with masks. Muffled speech compounded by the noise of a radio may easily cause a medication error.

The bottom line is that any kind of distraction in the OR can lead to error, and music can be distracting. Even if a person likes the music, he or she can be momentarily diverted by a favorite song or artist.[8]

may be helpful. For example, the staff may want to turn down the volume on alarms or use quieter medical carts.

■ Educate the staff on the importance of noise reduction. By using calm voices when communicating and considering the noise they make when going through their work, staff members can help further reduce noise in the OR. (See Chapter 1.)

■ Stop playing music in the OR. While the Joint Commission has no official position about playing music in the OR, if it interferes with safe patient care, organizations should consider eliminating it. (See Sidebar A-1.)

Ensuring the Safety of Compressed Gas Cylinders

Within every OR, there is a container that could act as a deadly weapon. The compressed gas cylinder houses necessary medical gas, such as oxygen, nitrogen, or nitrous oxide, but if inappropriately connected or stored, this canister can cause death and destruction.

Preventing Medical Gas Mix-ups

Several years ago, the issue of medical gas mix-ups caught the attention of the media and health care agencies, such as the Joint Commission and the Food and Drug Administration (FDA). Several patients who were supposed to receive medical-grade oxygen instead received different gases. The root cause analyses of these sentinel events revealed essentially the same thing: The organizations had failed to implement proper training.

Medical-grade gases, used either in medical procedures or to power medical equipment, include oxygen, compressed air, carbon dioxide, helium, nitrogen, and nitrous oxide. A limited number of people, mainly maintenance staff and respiratory therapists, are usually authorized to handle compressed gas containers, which feed into the piped gas system from a central location. An index system is used to ensure that the outlet fitting on the vessel will line up exclusively with the fitting on the manifold carrying the same gas so that the oxygen vessel connects to the oxygen system, the nitrogen vessel connects to the nitrogen system, and so forth.

Many medical gas containers are color-coded. Color-coding, designed to make a green vessel, for example, instantly recognizable as oxygen instead of nitrogen, is not always standard in the industry. Indeed, the Compressed Gas Association (CGA), a safety and standards organization of manufacturers and distributors, believes that the use of color-coding is a source of confusion. As the CGA points out, there are more than 100 compressed gases transported in the United States, but there are only 7 colors, plus black and white, used on tanks.

Adding to the problem are the discoloration that can result from daily wear-and-tear, inadequate lighting in storage units, and color-blindness.

Although medical gas cylinders also have an indexing system, it is a mistake, according to ECRI, an independent health services research agency, to rely completely on the connector's indexing system to positively identify the gas inside the vessel. Many connectors are removable and replaceable. Removal may require a wrench and a pair of pliers, but these are easy to come by in health care organizations. In one of the cases involving patient death, it was not even necessary to replace the connector: One of the indexing pins in the oxygen flow meter was missing, making it possible to fit the defective oxygen device into a nitrous oxide outlet.

For all these reasons, both ECRI and the American Society for Healthcare Engineering (ASHE) say that labels are an extremely important way to identify a gas cylinder's contents. If the label is missing or unreadable, ASHE's advice is simple: Do not use the cylinder. Period.

No matter what engineering controls are in place, the best way to prevent a medical gas mix-up up is through proper training. Everyone who handles medical gas should be taught to read labels carefully, to understand the purpose and use of differently configured cylinder and outlet fittings, and to appreciate the lethal implications of ignoring the one or tampering with the other. Toward this end, the FDA recommends the following:

- All personnel who will be handling medical gases should be trained to recognize the various medical gas labels and to examine them carefully before hooking connectors to the system.
- If the supplier uses 360-degree wrap-around labels to designate medical oxygen, personnel should be specifically trained to make sure each vessel they connect to the oxygen system bears such a label.
- All personnel who are responsible for changing or installing cryogenic medical gas vessels should be taught how–and why–to connect them properly.
- Training should emphasize that the fittings must not be changed under any circumstances. If a vessel fitting does not seem to connect to the oxygen supply system fitting, the supplier should be contacted immediately and the vessel returned so that the problem can be identified and corrected.

☞ **Tip:** When training the staff on how to effectively install medical gas cylinders, organizations should include hands-on training. Labeling and connector systems should be demonstrated, not just described. ☜

Other recommendations to help prevent medical gas mix-ups include the following:

- Store medical-grade and industrial-grade gases separately, in well-defined areas; full and empty cryogenic vessels for medical gas should likewise be kept separate.
- Check and document the container and connection once a cryogenic vessel is connected to the oxygen supply system but *before* gas is introduced into the system.
- Reject vessels from a supplier without written documentation of the concentration of gas in the vessel.
- Visually inspect each vessel upon receipt, before it is placed by the supplier, to ensure that an oxygen label is clearly marked and that no other conflicting labels or stickers are present.
- Make sure lighting in the medical gas storage area is adequate to read labels.
- Inspect all incoming devices to determine whether they have the appropriate pin indexes or inlet connectors. If a supplier's cylinders do not conform to national standards, ECRI advises to find another supplier.

Addressing the Physical Hazards of a Compressed Gas Cylinder

A typical gas cylinder is nearly five feet tall. It weighs more than 150 pounds when filled and is pressurized at 2,200 pounds per square inch. Compressed gas cylinders are typically stored upright and secured so they cannot topple over. However, some organizations bypass this storage option and rest them on a small platform with no other means of support. If a medical gas cylinder falls over and dislodges its cap, or if the cap is removed and the tank gets significantly jostled or bumped, all the tank's pressurized power will escape through an opening no bigger than a pencil, causing the cylinder to act like a missile. Some cylinders have been known to fly through the air, smash brick walls, or ricochet around a room. If a person or structure is in the way of the cylinder's path, severe consequences can result.[9]

To prevent gas cylinders from causing damage, organizations must properly store them. They must be secured so they cannot be tipped, bumped, or pushed over, typically in a rack. Even an empty container houses 100 pounds of pressure and can cause serious damage if dropped. Consequently, when storing empty containers for disposal, or to be picked up by the manufacturer, they too must be stored in the upright and secured position. Caps must also be securely fashioned so they cannot slip off.

☞ **TIP:** When transporting compressed gas cylinders, they should not be rolled but carried like a baby. If an organization such as an ambulatory care facility must transport a container to an outlying clinic, the cylinder should not be put in the trunk of a car, but rather stored in a rack and transported in a vehicle in which the rack remains upright. ☜

Standard EC.1.20. The organization maintains a safe environment.

Ensuring Laser Safety

The advent of lasers in medicine has helped revolutionize some surgical procedures. Lasers have made invasive procedures less so. Despite all the benefits of lasers, these tools emit radiation, and that radiation can have negative consequences on the safety of health care workers. For example, exposure to radiation at certain frequencies for long periods can lead to eye-performance issues, dermatologic hazards, and some forms of cancer. In addition, when used inappropriately, lasers can cause destruction and injury.

Following are some suggestions organizations can use to help ensure the safe use of lasers within the OR:

- Require proper eye protection for individuals using lasers. This protection differs depending on the type of laser used. Organizations should check the Occupational Safety and Health Administration (OSHA) Web site (http://www.osha.gov) for details on the appropriate eye protection.
- Identify and monitor individuals who may be at special risk from the ultraviolet hazards associated with laser use.
- Conduct baseline eye exams. The American National Standards Institute (ANSI) recommends that organizations conduct baseline eye exams on all the staff who will be regularly exposed to laser radiation. Those at risk may include surgeons, scrub nurses, anesthesiologists, and anyone else who is regularly around an active laser. This baseline exam helps determine an initial level of visual performance against which damage can be assessed in the event of a laser accident. ANSI recommends that the baseline exam be performed at a preemployment physical and another exam be performed in the event of an injury.
- Require the staff to wear appropriate respiratory protection.
- Use smoke evacuator units. As mentioned in Chapter 5, page 61, these can evacuate smoke from any laser plumes.
- Establish procedures for how to respond and treat an individual who has had harmful exposure to laser radiation, plume smoke, and so forth.

In addition to the health hazards associated with lasers, organizations should be aware of the dangers that exist when a laser is not used properly. Because a laser emits a highly powerful beam of radiation, it can be quite harmful if it accidentally comes in contact with a person or an object. Lasers can cause burns, electric shocks, and possibly blindness, if directed inappropriately. Lasers should never be aimed at someone or something and should be turned off when not in use. Any staff member who uses a laser should receive proper education on its use and the hazards associated with it.

☞ **TIP:** When installing laser equipment, organizations should make sure that the laser aims at the wall and not the door. If directed at a door that is unintentionally opened while a person walks in front of it, the laser may cause serious damage. ☜

☞ **TIP:** If users of lasers, electrosurgical equipment, power tools, and the like spill a substance such as saline on the equipment, users should unplug the equipment and send it to maintenance to be fixed. Continued use of the equipment is hazardous, and many times the equipment is easier to repair if it has not been used after a spill. ☜

Ensuring Anesthesia Equipment Safety

For a number of years, anesthesiologists have been aware that working around anesthetic gases, such as nitrous oxide, can present certain health hazards, such as reproductive issues, liver and kidney disease, and neurological disorders.

When delivering anesthesia to patients, more gas is used than is strictly necessary. Any excess gas is vented from the patient's breathing circuit through a pressure release valve. This valve prevents pressure from building up within the circuit. Unless this excess gas is removed, it can accumulate in the air of the OR and cause harm.[10]

The process of removing excess anesthesia gases from the OR environment is called *scavenging.* The primary objective of scavenging is to capture the excess gases released from the pressure relief valve and direct them to a waste gas evacuation system. Once in the system, the gases are then exhausted outside the building.

A scavenging system has three major components: a gas-capturing assembly that traps the gas for disposal, a disposal system that vents the gas outside, and an interface between the gas-capturing assembly and the disposal system. There are many different varieties of scavenging systems. Organizations should review the different types and select the kind that best serves the needs of the organization.

Even an effective scavenging system does not dispose of all anesthetic gases present in the OR. Excess gases can come from leaks, inappropriate use of the anesthesia equipment, and other equipment, such as a mass spectrometer. Excess gas from these sources cannot be captured by a scavenging system. Consequently, additional measures should be taken to ensure the most healthful environment possible and may include the following:

- Routine servicing and preventive maintenance of the equipment
- Purchasing low-leakage anesthetic equipment
- Using low-leakage anesthetic techniques. This includes turning on the nitrous oxide only after the face mask has been applied to the face of the patient.
- Conducting regular leak testing. This can be done in a variety of ways, such as by using an infrared nitrous oxide analyzer or by submerging the suspected leak site in water. If submersion is not possible, application of an 80% water and 20% liquid soap solution will expose any leaks.
- Regularly monitoring trace gases. This can be done through air sampling and testing or through the use of infrared analyzers.

Employees working in the OR should be regularly informed about the risks and possible toxic affects of exposure to anesthetics. In addition, in the preemployment physical, the organization should document the worker's baseline health, including any abnormal pregnancies. All records of such conditions should be maintained.

Ensuring Radiation Safety

Radiation is a key element in the modern OR. It can be found in lasers, x-rays, fluoroscopy equipment, and some radioactive materials, such as implants. There are two types of radiation: ionizing and non-ionizing radiation. Ionizing radiation is a low-frequency form of radiation that can be found in microwaves, radios, and lasers. Non-ionizing radiation includes higher-frequency radiation, such as that found in gamma rays and x-rays. Both types of radiation can be found in the OR.

Like other aspects of the OR environment, radiation can be dangerous. Depending on the level of exposure, radiation can pose a health risk to workers and patients. It can adversely affect individuals who are directly exposed as well as their descendants. Radiation can affect cells of the body, increasing the risk of cancer or harmful genetic mutations that can be passed on to future generations; or, if the dosage is large enough to cause massive tissue damage, it may lead to death within a few weeks of exposure.[11]

Perioperative staff can be exposed to radiation in a variety of ways, including through inhalation, absorption, or injection. Because perioperative staff members are typically clustered together around a patient, a lot of people may be exposed to a radiation source. To reduce the risks associated with radiation, organizations should take the following precautions:

- Radioactive materials, such as implants, should be left in their lead-shielded containers for as long as possible.
- The staff should use gloves when handling radioactive materials.
- The staff should wear protective masks and scrubs when working near radioactive materials that could be inhaled.
- The staff should wear eye protection when working directly with highly radioactive material.
- Lead curtains should be used when high-frequency radiation, such as x-rays, is needed. For example, when taking an x-ray to check for the presence of a missing sponge, a lead curtain should be used. If for some reason a perioperative staff member must place his or her hand behind the lead curtain, lead gloves should be worn.
- Individuals should keep a safe distance from any necessary shielding.
- Access to areas where ionizing radiation is used should be controlled.

Depending on the type of procedures performed at a particular facility, some organizations have a higher risk for radiation exposure than others. Organizations should determine their risks and put in place safety processes to address those risks.

Several health care agencies provide specific recommendations regarding radiation safety, including OSHA, the CDC, AORN, and AAMI. Organizations should consider consulting these safety agencies when developing radiation safety policies.

Addressing Ergonomic Issues

Ergonomics is the science of fitting a job to a worker. It involves designing equipment and work tasks to conform to the capability of the worker, and it provides a means for adjusting the work environment and work practices to prevent injuries before they occur.[12]

One of the primary ergonomic issues associated with the OR environment is the possibility for trips, slips, and falls. Floors that are slippery from blood and bodily fluid can cause an ergonomic hazard. Crowded work areas where staff members are moving at a rapid pace can also lead to possible ergonomic incidents. To help prevent these types of events in the OR, organizations should consider the following:

- Using nonslip surfaces in the OR
- Maintaining drainage and providing dry standing places where practical
- Providing appropriate waterproof footgear to surgical staff
- Scheduling the OR so that staff members are not rushed to complete a procedure but can walk calmly throughout the OR
- Educating staff members on how they can reduce their likelihood of slips, trips, and falls in the OR
- Posting signs that remind staff to slow down and be careful
- Encouraging staff to recognize that even if the floor does not look wet, it may be. For example, if someone has just spilled saline on the floor, it may not be apparent, and a slippery situation could be present.

☞ **Tip:** When configuring an OR, an organization should design it to minimize possible trip and fall hazards. ORs should not be crowded with equipment or difficult to navigate. ☜

Another ergonomic risk in the OR involves lifting. Patients must be lifted on and off the operative table, and improper lifting strategies can result in injury. To lift properly, the staff should be educated on how to lift safely. Some things the staff should know when lifting include the following:

- Never transfer patients when off balance.
- Lift loads close to the body.
- Never lift alone. Use team lifts or use mechanical assistance.
- Avoid heavy lifting, especially when the spine is rotated.

Organizations should limit the number of allowed lifts per worker per day and train staff in when and how to use mechanical assistance.[12]

References

1. Joint Commission Resources: How to improve information flow: Ensure communication between plant and clinical personnel. *Environment of Care News* 6:11, Jul. 2003.
2. Occupational exposure to bloodborne pathogens: Needlestick and other sharps injuries: Final rule. *Fed Regist* 66, Jan. 18, 2001, pp. 5318–5325.
3. Hospitals for a Healthy Environment: *Regulated Medical Waste Reduction.* http://www.h2e-online.org/docs/h2e10steprmw20103.pdf (accessed May 5, 2006).
4. Blanchard J.: *Clinical Issues: Infectious Waste Disposal.* http://www.aorn.org/journal/2003/octci.htm#6 (accessed May 5, 2006).
5. U.S. Department of Transportation: *Infectious Substances: What You Need to Know, Regulated Medical Waste Packaging Hazardous Materials Safety Brochure,* Pub. no. 49 CFR 173.197. Atlanta: U.S. Government Printing Office, 1997.
6. Conner R.: *Clinical Issues: Copies of Surgery Schedules.* http://www.aorn.org/journal/2001/julyci.htm (accessed May 6, 2006).
7. Smetzer J.: Prescription for safety: Cutting errors out of the operating room. *American Hospital Association News* Jun. 3, 2002, p. 6.
8. Fogg D.: *Clinical Issues: OR Music.* http://www.aorn.org/journal/2002/novci.htm (accessed May 10, 2006).
9. Magnusson N.: *Compressed Gas Cylinders: The Physical Hazard.* http://safety.science.tamu.edu/cylinderhazard.html (accessed May 8, 2006).
10. Blanks S., et al.: Scavenging of waste anesthetic gases. *PTSM* 1987, pp. 27–33.
11. Environmental Protection Agency: *Ionizing Radiation Fact Sheet Series: No. 1.* http://www.epa.gov/rpdweb00/docs/ionize/402-f-98-009.htm (accessed May 10, 2006).
12. Occupational Safety and Health Administration: *Ergonomics.* http://www.osha.gov/SLTC/etools/hospital/hazards/ergo/ergo.html (accessed May 12, 2006).

Appendix 2: Maintaining a Drug-Free Workplace[1]

A surgeon's hand trembles before surgery. He or she seems very moody and distracted. A nurse seems increasingly distracted and even dysfunctional. An anesthesiologist falls asleep in the middle of a procedure. Among the many possible explanations for such phenomena is worker substance abuse, a serious health problem that can have devastating effects in the operating room (OR) and health care environment in general.

In fact, substance abuse is one of the nation's largest health problems. Although many stereotypes exist about alcoholics and drug users, the truth is that addiction knows no boundaries, and people who struggle with addiction come from all walks of life. Most, in fact, are employed, and no industry–including health care–is immune.

The Occupational Safety and Health Administration (OSHA) and the Joint Commission have been working in partnership to improve the safety of health care workers; the Joint Commission is always concerned with the safety and well-being of patients. Increasing the understanding of substance abuse in the health care environment is important because such abuse can compromise the well-being of both patients and workers.

"OSHA's mission is to ensure a safe and healthy working environment for America's workers," says Don Wright, M.D., M.P.H., director of OSHA's Office of Occupational Medicine. "OSHA recognizes that impairment by drugs or alcohol can constitute an avoidable workplace hazard, and we encourage organizations to implement drug-free workplace programs to reduce this hazard."

Joint Commission Standards

The need for programs to reduce substance abuse is reflected in the Joint Commission's standards. Element of Performance (EP) 5 of standard E.C.1.10 specifies that the organization uses the identified risks to select and implement procedures and controls to achieve the lowest potential for adverse impact on the safety and health of patients, staff, and others coming to the organization's facilities. "A preemployment or for-cause drug-testing program is important for all staff to ensure safety," says A. Kenneth Peterson, M.H.A., F.A.C.H.E, associate director, Standards Interpretation Group, Joint Commission. "For instance, it would hardly be safe for impaired workers to operate motor vehicles or machinery in the course of their work."

Standard MS.4.80 requires all accredited hospitals and medical practices to have a process in place that addresses health issues of licensed independent practitioners in a way that is separate from discipline. That process must be confidential and must enable licensed independent practitioners to come forward and be referred for care. In addition, a health care organization must have an educational process that addresses the health of licensed independent practitioners.

"There has to be a process to handle issues related to the health of the licensed independent practitioner," says John Herringer, associate director, Standards Interpretation Group, Joint Commission. "Organizations are sometimes

Sidebar A-1. Department of Labor Program for a Drug-Free Workplace

The U.S. Department of Labor's Working Partners program helps employers and employees work together to ensure a safe and drug-free working environment. Its Web site, http://www.dol.gov/workingpartners, offers an interactive tool to help organizations develop a drug-free workplace policy, along with a range of training and educational materials for use in the workplace. "A sound, well-executed policy can provide the critical first step to treatment and recovery," said Elena Carr, the department's drug policy coordinator. "It can also protect the safety and health of workers and the public."

reluctant to ask questions when they're concerned about the health of a licensed independent practitioner. The cause for concern could be a wide variety of health issues, such as alcohol on the breath or hands with a noticeable tremor."

Department of Labor Program

The U.S. Department of Labor's Working Partners for an Alcohol- and Drug-Free Workplace program (see Sidebar A-1, above) works closely with OSHA to help organizations develop and implement drug-free workplace programs. A comprehensive drug-free workplace program generally includes five components: a policy, supervisor training, employee education, employee assistance, and drug testing. Elena Carr, the department's drug policy coordinator, urges organizations to choose any or all of these components, selecting those that make sense for their workplaces. "Any policy should make it clear that the organization won't tolerate any use of illicit drugs or abuse of legal drugs, including alcohol, over-the-counter drugs, and prescription drugs," she says. "It should also emphasize that help is available for those who need it." According to Carr, when drug-free workplace programs include early intervention, access to treatment, and follow-up services, they can save the lives and livelihoods of health care workers while keeping patients safe.

Physician Health Programs

Some hospitals develop their own staff health programs. In addition, there is a widespread physician health movement to address openly the issues of physicians' health and the possibility of impairment. One example is Physician Health Services, Inc. (PHS), a corporation of the Massachusetts Medical Society in Waltham, Massachusetts. The director of that program is Luis T. Sanchez, M.D., a psychiatrist with additional certification in addictions. "Virtually every state has something similar to what we have in Massachusetts," says Sanchez. More than 40 states belong to the Federation of State Physician Health Programs, of which Sanchez is the president. "Most of these programs are available to all physicians through self-referral or a referral from a colleague, a family member, or the chief of service when it looks as if a physician might be developing a problem of substance abuse, often from alcohol or self-prescribed medications."

Testing for Health

"The use of illicit drugs such as marijuana and cocaine is not high on the list of physicians' problems," reports Sanchez. He states that alcohol abuse is more common, as is the misuse of opiates first prescribed for maladies, such as back pain, but then self-administered by physicians to treat their own symptoms. "Our programs are highly confidential," Sanchez says. "If we believe the physician could benefit from being monitored, we initiate a formal contract that allows us to screen the physician for substance use during the subsequent three to five years." Those tests may involve urine screens or testing of the physician's hair or blood.

Sanchez reports that the range of drugs for which physicians are tested and the allowable limits are far more stringent than even for airline pilots, truck drivers, or bus drivers. "We want to see abstinence, not just low levels of marijuana, cocaine, or opiates. We insist on total safety." And instead of allowing physicians to treat themselves, PHS refers participants to community psychiatrists and therapists for treatment.

The program also relies on the services of workplace monitors. "If a physician is well enough to work, we make sure that at least two physicians in the workplace know the physician is participating in the program and are aware of the problem," says Sanchez. These monitoring physicians report back periodically on how the monitored physician is doing. The monitored physician is usually required to attend a specified number of meetings of a self-help support group.

Sanchez stresses the importance of prevention, which means trying to reach medical students, residents, and physicians *before* they get in trouble. "We try to show them the folly of, say, having a glass of wine while they're on call." Most states have an incentive for physicians to get help *early.* "Our problem as physicians is that we don't want anyone to know there's something we're struggling with," says Sanchez. "While that's true of most people, doctors see themselves as healers. And patients who look to their doctor for healing don't want to know that their doctor has a problem. We're trained to take care of others, rather than ourselves, and that's really unfortunate. But once a physician recognizes what's at stake with respect to patient safety and his or her own licensure, the outcomes are usually very good."

A Case Study: Vanderbilt University Medical Center[2]

Vanderbilt University Medical Center is an 800-bed tertiary care hospital that incorporates Vanderbilt School of Medicine and the Vanderbilt School of Nursing. Vanderbilt Clinic had more than 698,900 patient visits during 2003, and Vanderbilt Hospital admitted more than 33,800 patients.

By 1998, it had become apparent that the health of clinicians and others on the staff of Vanderbilt Hospital needed attention. In the years preceding 1998, the careers of eight physicians were cut short by tragedy. Four physicians died: two were suicides, one died from domestic violence, and one died from an indeterminate cause. Four other incidents, including three involving sexual misconduct, also came to the attention of Mary Yarbrough, M.D., M.P.H., director of Health and Wellness at Vanderbilt University and director of the Vanderbilt Center for Occupational and Environmental Medicine.

Sidebar A-2. How Widespread Is the Problem?

According to the National Survey on Drug Use and Health, sponsored by the Substance Abuse and Mental Health Services Administration (SAMHSA), during 2003 and 2004, about 5.8% of all health care practitioners and health care technical personnel reported using illicit drugs (including marijuana/hashish, cocaine [including crack], heroin, hallucinogens, inhalants, or prescription-type psychotherapeutics used non-medically) in the previous month.[3] Moreover, about 1.1% reported dependence on or abuse of illicit drugs in the previous year. More alarming still is that about 5.3% reported dependence on or abuse of alcohol in the previous year. For example, among physicians and surgeons who diagnose and treat patients, about 4% reported using illicit drugs in the previous month, about 0.8% reported dependence on or abuse of illicit drugs in the previous year, and about 4.4% reported dependence on or abuse of alcohol in the previous year. In addition, about 7.4% of nursing, psychiatric, and home health aides reported using illicit drugs in the previous month, about 1.7% reported dependence on or abuse of illicit drugs in the previous year, and about 6.1% reported dependence on or abuse of alcohol in the previous year.

In response, members of the Vanderbilt Medical Board agreed to initiate a physician wellness program focused on preventing impairment and educating physicians who were already impaired, whether from mental health issues or drug or alcohol abuse. The Vanderbilt program preceded the Joint Commission standard on licensed independent practitioner wellness but matched it closely and included many of the same components specified in standard MS.4.80–education, prevention, referral, diagnosis, treatment, evaluation, reporting, and monitoring.

Dr. Yarbrough describes the programs–the Physician Wellness Program and the Nursing Wellness Program–as subsets of Vanderbilt's Work/Life Connections Employee Assistance Program (EAP). "Our programs cover not just problems with alcohol and drugs but all mental health concerns, and the majority of cases are self-referrals," says Yarbrough.

When impaired medical professionals enter the Vanderbilt program as either self-referrals or mandatory referrals (treated the same as self-referrals), they are evaluated, diagnosed, and allowed to choose from among three of the community's best providers with experience in physician impairment. The Vanderbilt Center works with the physician's insurance company and assigns a staff psychiatrist as well as a social worker who functions as a case manager until the physician recovers and/or the case is closed.

"We help them identify whatever resources they need," says Yarbrough. "We follow them throughout the program, and if they're mandatory referrals, we follow them for at least six months afterward. If they're getting help for an addiction, we might follow them for as long as three years." Yarbrough points out that if a physician disagrees with the need for treatment, as is often the case with disruptive behavior, Vanderbilt may send the individual elsewhere for an independent evaluation. "The main thing," says Yarbrough, "is to get a diagnosis and a treatment plan."

In addition to the physician and nursing wellness programs, Vanderbilt also offers help to members of its staff and faculty. "The priority in every case is patient safety," says Yarbrough. "We ask a number of questions: Is it safe for this person to work? Are they a threat to themselves or anyone else? Do they need any accommodations for their impairment?" Yarbrough deals with organizational issues such as the following: Why is this referral mandatory? Are there any disciplinary issues? If so, have they been dealt with in the appropriate format? Are there any legal issues or employee-discrimination or employee-relations issues? Yarbrough also collaborates with other department heads to make sure these concerns are handled according to Vanderbilt policy. "With mandatory referrals, we bring in the physician and explain how the program works, what is considered confidential, and when information is released," says Yarbrough. Unless the physician is a threat to patient safety or the law requires otherwise, a release for any information that is distributed about that physician must be signed by the physician himself or herself. All such information is strictly on a need-to-know basis.

Meanwhile, Vanderbilt psychiatrists and social workers continue dealing with the clinical components of the person's recovery, such as the following: What is the diagnosis and treatment plan? Is it safe for the person to practice? Are there any restrictions to his or her practicing? Is any monitoring needed in the workplace?

Changing Parameters

When Vanderbilt launched its program six years ago, it was helping no more than eight people a year. Most of them had already started treatment for drug and alcohol problems and were coming to the Work/Life Connections EAP for drug screenings. Since then, the number of people treated in the program has increased substantially. Between April 2005 and January 2006, Vanderbilt's Physician Wellness Program received 120 referrals, of which 90% were self-referrals. Only 10% were for drug and/or alcohol addiction; the remaining referrals were for other issues.

Self-medication with a prescription drug is often unmasked in the course of working with other behavioral health conditions. Yarbrough points out that depression and other psychological problems are major concerns at present, accounting for 40% of referrals. "There is a whole other group whose primary problem might not be addiction, but in the course of the recovery, they'll be treated for that as well," she reports. Thanks to extensive experience, the professionals who run the Work/Life Connections program are able to identify problems early and deal with them before they become acute and involve disciplinary issues. "The majority of self-referrals are simply unhappy," says Yarbrough, "and they want help for their illness."

Areas of Concern

According to Yarbrough, anesthesiology is an area of concern because of easy access to prescription drugs. "There's also the abuse of alcohol across the board," she continues. Yarbrough has found that, with physicians, unless they're very much out of control, their impairment can be quite difficult to detect. It may come down to a dramatic event in the workplace, and sometimes several incidents have happened well before that. "The more referrals we get, the earlier we're likely to pick up on alcohol and drug addiction," says Yarbrough.

Of the 120 people who are treated in a year, about 40% are seen for psychological problems, 20% for relational and family issues, and about 30% for workplace behavior, such as being disruptive or acting irrationally. Finally, about 10% are seen for alcohol and drug problems, with the remainder considered "other."

To care for the health care providers, Vanderbilt has one psychologist, four social workers, three psychiatrists who work part time, and a nurse practitioner who works with the Nursing Wellness Program.

As noted, the Vanderbilt program has expanded beyond physicians to cover members of the faculty, the nursing staff, and also the postdoctoral population at the medical center. Yarbrough reports that the faculty and staff members who are served by Vanderbilt's Physician Wellness Program number about 2,500; these individuals are mostly clinical and include residents. "We're finding increasing numbers of special needs among professionals," she reports, "because of their licensure requirements and because they're doing clinical work involving patient safety issues. The process is far more complicated when it involves licensure."

Yarbrough attributes the credit for the program's success to the Vanderbilt medical staff. "The leadership here has been quite supportive. They want to reduce stress and burnout among physicians. Instead of 'catching' and 'punishing' impaired clinicians, they understand these behavioral mental health and substance-related conditions to be diseases and, as such, to require appropriate diagnosis, treatment, and follow-up. With patient safety as the number-one priority, the issue is, how can we help the physicians?"

The message of the Vanderbilt program is that, like everyone else, clinicians have mental health and substance abuse problems. The majority of self-referrals respond well to appropriate referral for counseling and treatment. Those clinicians who go to treatment for addiction or who are especially disruptive must sign a contract with multiple components. They must also check in periodically with the Vanderbilt program, in which they are subject to random drug screenings.

Those with addictions must also enter into an agreement with the state of Tennessee. "When they

Sidebar A-3. Issues Addressed by Vanderbilt's Work/Life Connections Program

- Stress/emotional issues
- Depression/grief/loss
- Family/marital conflicts
- Interpersonal concerns
- Divorce adjustment
- Domestic violence
- Alcohol/substance abuse
- Relationship issues
- Critical-incident stress management

get their health back, and their disease is in remission," says Yarbrough, "we continue to support their recovery and help them get back to work. They have a big financial, intellectual, and emotional investment in their professional capability. And more than 95% of the referrals who require an agreement make it through our three-year program."

References

1. Joint Commission Resources: Helping drug-impaired health care workers. *Environment of Care News* 6:1, Jun. 2006.

2. Joint Commission Resources: Workers get help for drug, alcohol, other impairment: Vanderbilt University Hospital runs successful treatment programs. *Environment of Care News* 6:6, Jun. 2006.

3. Substance Abuse and Mental Health Services Administration (SAMHSA): National Survey on Drug Use and Health, 2003 and 2004. Rockville, MD: SAMHSA.

INDEX

A

B

C

D

E

F

N

O

P

Q

R

S

T

U

V

W

X

Z